POWDER DAYS

Also by Heather Hansman

Downriver: Into the Future of Water in the West

HEATHER HANSMAN

POWDER DAYS

SKI **BUMS,**

SKI **TOWNS**

AND **THE FUTURE**

OF CHASING **SNOW**

HANOVER
SQUARE
PRESS

HANOVER
SQUARE
PRESS™

Recycling programs
for this product may
not exist in your area.

ISBN-13: 978-1-335-08111-7

Powder Days: Ski Bums, Ski Towns and the Future of Chasing Snow

This edition published by arrangement with Harlequin Books S.A.

Hanover Square Press
22 Adelaide St. West, 41st Floor
Toronto, Ontario M5H 4E3, Canada
HanoverSqPress.com
BookClubbish.com

Printed in U.S.A.

TABLE OF CONTENTS

INTRODUCTION: THE DREAM

We duck our heads into the wind, ski tips skittering as we curve along the ridge, slowing down to scout the gullies and pick our lines.

Katie drops in first, and from above I watch her turn, every arc short but smooth. She sends down a wave of sluff as she carves through the chute, whooping, more graceful than I'll ever be. In the apron she stops below a knuckle of rock—out of the line of fire if I were to fall or if the slope were to slide—and waves me down. I lean forward in my boots and tilt over the edge, giving in to gravity.

This rib of rocky chutes was out of bounds when we ski patrolled here at Arapahoe Basin, almost a decade ago. Back then, the only way into the Steep Gullies was to wait for conditions to line up, then duck a rope and sneak around the elbow of the ridge, hoping no one saw you go, hoping that the sugary Colorado snowpack would be stable when you did. I followed older guys past the boundaries, nervous but glad to be invited, hopped up on a mixture of adrenaline, secrets, and the thrill of breaking rules.

Now the gullies are in-bounds, kicked open by the pressure of resort expansion and skiers hungry for terrain. From the ridge you can see down to Highway 6, where the real dirtbag skiers used to park their campers, sleeping close to the hill, cutting corners wherever they could. These days traffic stacks up in the canyon, and you're not allowed to sleep in your vehicle anymore. It feels like all of Denver drives to the mountains every weekend, cramming the slopes, skiing off every bit of snow. It seems, sometimes, like there are no more secrets.

Part of that is true. It has changed. It is crowded, more expensive, and more exposed. But my frustration is also driven by the shifting baseline of nostalgia, and the way the past resonates and echoes when I'm back in a place that used to be mine.

Even if things had remained the same, we would be different. Katie has a kid now, and I live seven states away. We no longer subsist on quesadillas and High Life, and we're back here in passing. We're not locals anymore—even when I still try to hold on to those dirtbag secrets and claim them as mine.

But we *were* ski bums once. Or at least we were skiers chasing the idea of wildness, helixed around an obsession with untouched snow and the kinds of parties where someone often ended up naked, shooting fireworks out of their ass. For a while we were insiders, stratified into the mountain town ecosystem, shiny in our youth and ease, even when we were serving pizza second shift to supplement our skimpy on-mountain paychecks. Even when we were stamping our feet in the bitter cold, scanning lift tickets while some sweaty second home owner accused us of ruining his vacation.

We were living the "dream," prioritizing skiing over everything else. Heavy on the quote unquote. We moved to the mountains and let the other facets of our lives fall into place from there. It was an idea I'd been obsessed with since I was a kid when I first felt the adrenal glory of downhill motion, lit from behind by the perfect sunset shots of ski movies and glossed up by magazines. I thought that being a skier—a real skier—meant committing to the constant sense of chase.

★ ★ ★

Now I'm not sure if I remember the beginning correctly, or if it's become a story I tell myself, one of those light-on-the-threshold moments. But here's what I remember: there was a campfire and we were in Maine. Ivan, my future boss, who was a semi-stranger then, might have been wearing a mullet wig, as he often did that summer. We were in the thin ellipsis between bug season and winter and I was just over the edge of twenty-one, just out of college. Not sure what was coming next, but sunk into that achy-chested, pre-melancholy of knowing I'd never be back there, or at least not in the same way. I was raft guiding on the granite-choked rivers of the northern woods, sleeping crammed into a two-room, eight-bunk cabin with at least nine other people, living on leftover trail mix. I had already, unwittingly, primed myself for dirtbag living by neglecting showers, counting my income in tips, committing to moving my body hard.

When Ivan said he could get me a job in Colorado, where all I would really have to do was ski every day, my life pivoted toward a particular grimy dream. I don't believe in God or fate, but some tangled part of me got sucked into a modern manifestation of the frontier fantasy, problematic as it might be. I latched on to the idea that if I went west I would be braver and truer and more exciting. I wanted an adventure I could call my own, and a way to grow up with the country. A path that feels hard to find now that so much is commodified and mapped. I just needed someone to tell me it could be real.

So yes, I moved halfway across the country to a town I'd never been to before because a man in a wig said he could get me a job scanning lift tickets for minimum wage. And yes, I was a couple of beers deep when I said I would go.

The unknown was part of the appeal, but I told my mother I'd probably be back in May. I recruited two college friends to come with me, and drove to the Rockies in my decade-old Jetta.

Skinny East Coast skis stuck through the slot in the trunk, the rest of my life piled around them, chasing a conflagration of all the things I wanted to feel: the rip of gravity, independence, and interdependence, the adrenaline that comes from risk.

Of course, that wasn't the actual beginning. The actual beginning was a bunch of decades back, before a lot of snotty-nosed lift rides, broken fingers, and ski bus crushes brought me to that bonfire. You don't become a skier by accident—it's an objectively stupid, expensive, gear-intensive sport—but my parents enabled it early, cramming my brother and me into hand-me-down boots and carting us to New Hampshire, so they could ski, too. As a kid I found a sense of community tailing older boys down icy blue bumps. It made me feel different, and it made me feel like I belonged. In college, I'd wake up in the post-party, predawn dark to drive across Maine and New Hampshire just to ski knobby backcountry lines in the White Mountains. I've always felt clearer in motion.

My skiing story isn't unique. I'm a stereotype, an East Coast kid pulled west by the promise of bigger adventures and broader ranges. And I revered the notion of the ski bum. The idea that the purest skiers were the ones upending elitism in a world that was otherwise highly elite. Ski bums are the ones who operate below the surface, taking advantage of the system, but not exactly being in it. Never paying full price for food or skis or passes. That idea has long, tangly roots. As soon as lifts started turning, a counterculture that rejected social norms in favor of 100-day-a-year ski seasons sprung up around them. Shelled out by World War II, soldiers from the 10th Mountain Division came back to the US and started building ski lifts and cutting trails. They'd been groomed into iconoclasts who couldn't settle back into normal life, and they built up an industry around that drive. After that, Warren Miller glorified living in a parking lot in the '60s. The '70s made skiing look sexy and brave, and the '80s ushered in a hedonistic vision of sunny après and steep slopes. The romance of that story is strong. I am not the only one who ended up in the mountains on a loose-limbed bonfire promise. I've heard more stories of coin tosses and cars that happened to break down in Breckenridge than I have intentional, well-thought-out paths to life in a mountain town.

For me, it flipped into a fixation fast. Skiing threaded through everything from the jobs I chose to the people I loved. I spent that first winter believing that skiing could be enough. In May I signed another lease, and then another.

Like any obsession, the reality was more complicated. Even calling yourself a bum signals a level of privilege, and many people who claim the identity often come from money (and lots of it), which affords them their vagabond lifestyle in the first place. There's an imbalance that comes from chasing something fleeting and selfish because it can all fall apart in one bad fall. Scanning lift tickets sucks. Ski towns, and the built-up fantasyland of resorts can be havens for inequality and abuse. Skiers are drawn to excess and there is plenty to go around.

When I moved to Colorado in 2005, risk was everywhere, but in a way that felt exciting. I liked the brag of drinking too much, and I was too naïve to notice harder drugs. Climate change seemed vague and theoretical, and no one I knew had died in the mountains yet. The Great Recession hadn't shaken my generation's idea of stability yet and corporate entities were just starting to binge buy resorts. I didn't understand the direct connection between my rampant anxiety and my relentless impermanence, and I thought living in my car was cool. I thought I could exist like that forever.

But skiers know that a turn only looks elegant when it's in motion. It breaks down and loses its grace if you stop moving, and ski bumming is sort of the same. It exists in a liminal space, one that seems particularly fragile now, as resorts consolidate, climate change turns its screws, and economic inequality underscores existing disparities that concentrate in vacation

towns. Personally, politically, and ecologically skiing is changing. "A way of life is in jeopardy," Aspen's Vice President of
Sustainability, Auden Schendler, wrote in a recent climate report. It's hard to live forever in the bubble, and it's particularly
hard when it's squeezed by unreal expectations, and shrinking
spaces. Ski culture glorifies its iconoclasts. But that means that
as skiers we've been fed a narrow ideal that's nearly impossible
to achieve—which feels like a generational story about much
more than skiing.

For a dream that seems so bright, it is full of darkness. I saw
overdoses and overdrafts, and the kind of sadness that comes
from never feeling like you have enough. I know bodies and
brains broken by avalanches and untimely falls. Even the future
of winter itself is tenuous, and there's a fear that comes from
both holding on and letting go. You're selfishly pegging your
life to something beyond your control.

After a while I couldn't handle that risk. I got scared that
I'd wake up at fifty, still grumpily scanning lift tickets and be-
moaning the lack of snow. I split for nonseasonal work, and a
life in the foothills instead. I wasn't even thirty and I felt old.
But I didn't loosen my grip too much. I became a writer and
a ski magazine editor, hovering on the margin of the world I
used to embody—clinging to the edges. The irony is that from
the outside I was blasting the secrets I used to protect, trying to
profile the hardest-to-pin-down dirtbag in order to prove that
I was once an insider.

I know that life isn't easy, but I also know that obsession
is never rational, and that the allure of chasing motion is still
somewhere in my guts. Sometimes I think that I messed up by
abandoning the dream.

I see people I started ski bumming with still doing it. Last
winter, during a January storm, I walked into Pazzo's, the Vail
restaurant where I served late-night pizza and lived off shift

meals when I was twenty-two, and Zach was still standing behind the counter. Wrinkled and rounder, but still there, grin-
ning at some of the same barflies. When we slipped back into
the same conversations I wondered what my life would look
like if I had stayed, and how much of the story still holds true.

I'm still not exactly sure what pulls me back to the mountains
every winter, or why I've convinced myself that the best place
to be is on a wind-screwed ridgeline or drinking beers with my
gloves on by someone's tailgate. I don't know if it's people, place,
or the pursuit of something crystalline and quick-lived. The pay-
out comes in flashes: the G-force of a groomer rail, or the first
weightless drop over a cornice. But I do know that each winter,
as soon as the snow falls I'm shin-banged and shaky-thighed,
tracking SNOTEL gauges in the Sierra and the Sawtooths, driving hours through storms for a few downhill seconds, to chase
the perfect, fleeting weightlessness of powder turns.

Skiing, at its simplest, is the feeling of slipping past gravity.
Skiers chase snow and freedom and wildness, at the expense of
a lot of other things. I'm still trying to understand how something so ephemeral can shape your whole life. I know it's hard
to sustain—I couldn't—but I'm still obsessed with where that
dream came from, what makes it complicated, and why it still
exists. It's not just the gritty dirtbags, or the trust-fund ski bums
who are tied up in it. They're part of an ecosystem of skiing
which encompasses everyone from celebrity vacationers to the
undocumented lifties who enable their trips from Sun Valley to
suburban Massachusetts. This is the story of recreational skiing
in America—a pastime that's become a lifeblood for many—
and in the face of climate change, economic upheaval, and so
much more, I'm trying to figure out if skiing as we know it
will survive.

So last winter, a decade and a half after I first moved west, I
got back on the road. I crammed the necessary parts of my life

into my hatchback and drove to the mountains, fatter skis stuck in the trunk this time. I spent the season slung back into constant motion, chasing storms, crashing where I could, looking for true ski bums, to see if they still exist, and if they can continue to. I wanted to track down the past that brought me there, to see the current state of things, and the way it might fall out in the future. It was me chasing me chasing the dream, trying to understand my obsession.

That's why I'm following Katie into the Steep Gullies again, tracing the tracks of our old secrets. The sun slants down as we descend, pearling light across the snow like a wave. We link turns to the bottom of the slope and then try to hold our speed through the traverse where we once knew every drop and compression. We circle back to the Pali lift line, where I squint at faces I try to tell myself are familiar, even though I don't know any twenty-three-year-olds in this town anymore. Coming back cracks the fantasy, but it underlines it, too. My body knows the neurological pattern of fall line turns.

My heart gets sore, sometimes, for the lives I didn't live. I know that ambitions crack under the pressure of time, and bodies do, too. Capitalism is cold in the service industry. I've seen the ways my former friends struggle with getting older in these unforgiving towns, where leisure isn't easy, and comparisons are hard. But I ache for the good parts, too. Sometimes I get a glance at the slanted perversion of someone else's perfect-looking, powder-choked life. In that frozen moment, the fantastic myth of the forever ski bum clicks into place and I am back chasing that storm again.

SECTION 1

SKID LUXURY

Who is a ski bum

TRAM LINE

I show up early, but not early enough, caught in the dull star-dust of a low-pressure morning on the tail of a storm. I'm slow in the car, then late to the shuttle bus, slamming my boots on in the Stilson parking lot, shoving an extra pair of goggles, a packet of Pop-Tarts, and a couple of beers into my pack. A backpack with avalanche gear feels mandatory at Jackson Hole, where the stretch of the Tetons right outside the backcountry gates hover in the periphery.

The road from the town of Jackson Hole to the base of the mountain winds north along the edge of the Tetons, sharp peaks to the left, a sprawl of ranches in the valley to the right. When the bus finally pulls up to the base area, the line of skiers is already wrapped around the tram dock, even though the first tram won't ascend for another hour, and there are fat, soggy flakes in the air. Last night's storm came in warmer and skimp-ier than expected. It dumped nearly six inches of water-laden snow—less than forecasted—but there's still a line, because there's always a line. The guys in front in their black and tat-ter are studiously avoiding eye contact with anyone they don't know and nodding to those they do. AC/DC's "Thunderstruck" garbles out of the loudspeakers and there's a pent-up animal energy in the air. I stake my place in line, probably two trams out, and wait while the flakes build up on my shoulders and lashes.

Every ski town has its own feeling, and even though locals complain that Jackson's is now distorted through the swell of corporate pass holders and second home owners who use Wyoming as a tax haven, there's still a ski culture here, in the shaggy, committing peaks of the Tetons. Jackson is where dirtbag living, mountaineering, and the pushing of limits converge. The stories that shake out from this tram line have shaped what I think it means to be an obsessive skier, and what it means to be cool.

The first tram cranks through the bull wheel, and we all do the shuffle dance through the maze, gloved hands hooked on our ski bindings. Look quick, but don't be obvious, and you might notice the detail of a worn flannel, or an old patch, or next year's skis. You might snake past Duct Tape Pete, Buddha, or Wild Bill, legends in this particular lift line. The texture of the town, and the way it shaped ski history, stem from this lineup.

My group swells onto the tram dock, and then into the car. We ride up into the soup of the storm, hemmed in by coffee breath, sweaty Gore-Tex, and nervy energy, packed too close to move. When we swing at the last tower, close to the precipice of Rendezvous Bowl, the tram operator turns down the pulsing metal to give us a safety warning: be smart; bring avalanche gear; travel with partners; you can always ride back down the tram if you're scared. "And if you don't know…" he says, leading us into his last terrain warning. "…don't go," we all recite back, polite as kindergarteners. The doors click open and we shuffle into the gale.

You can't just call yourself a ski bum. Posturing gets called out quickly around here. Instead you dedicate yourself to something singular, undeterred by the societal pressure to get a job, or make money, or grow up. Only when you do it year after year, prioritizing skiing over everything else, do you become a ski bum. At its best bumming is ascetic and nature centered, physical and purposeful. At its worst, it's indulgent and immature, selfishly exclusive, locking those who do it in a perpetual state of unappealing adolescence. There's a saying, "no friends on a powder day," but it's more than that—no job, no relationship, no stability, nothing to work toward. But skiing can also foster deep connections, community, joy—and a healthy, heady rush that feels so, so good.

But it's hard to balance dirtbagging with the pressures of growing up. It's a question that keeps coming up in my own life, which still orbits around skiing. Can you be a ski bum forever? Would you want to?

I'm still skating around the edge of those questions: single, property free, untethered enough to ditch my life for a winter and get on the road. They're similar to the ones I've been asking myself since I moved to the mountains at twenty-one, for a winter that ended up lasting years. That first spring, after the snow stopped coming, I felt ashamed telling my college friends, who had health insurance, and the kind of boyfriends who would become husbands, that I was out of a job and unsure of where I was going to live for the next few months. But I've also, at turns unwittingly and intentionally, gravitated toward that instability. I love watching seasons roll in, and the instinct you acquire from paying attention to a snowpack over time. Antsiness is my most notable quality and I haven't quite aged out of it yet. So I'm here, tracking the anthropology of my own obsession.

Skiing's existential nag is its ephemerality. The fleetingness of snow. The feeling that it might have been better before you got there—that it *must* have been better before—because the stories make it seem wilder, more adventurous, and more free. And since skiing's past in the US really isn't that long, some of the people who shaped the stories are still skiing, especially here in the Tetons. I get a contact high when I look around. Lining up for the tram, I feel like I'm tapping into skiing's history, and also its potential future, which is why Jackson felt like the best place to start.

Later that morning, after I take a few runs off the tram, my phone buzzes with a text from Benny Wilson. Tram lap, backdoor, it says, and even though I'm high in the Hobacks, swerving through the almost-bumps, I beat myself back down the mountain to try to catch him, skiing sloppy in the rapidly warming chop. I find him just in time at the back of the tram dock, where ski instructors and patrollers are allowed to cut the line. "Pretend you're my client," he says, and we slip through, ahead of the crowd.

On the loading dock Benny pulls off his gloves to point out the dimensions of the skis he made himself. He has gnarled fingers and knobby hand-hewn rings, a loud gravelly voice you'd be able to pick out anywhere, and magnetic energy. People keep coming over to talk to him, and my questions keep getting interrupted. *Legend* is an easy word to throw around lightly, but Benny's Jackson Hole ski bum story has calcified into a myth.

Those people on the tram dock, with their air of knowing and their next-season gear, are carved in the image he created in the '80s, when he could get away with being a punk because he was the best skier on the mountain, and when he led a crew of hardcharging, no-fuck-given skiers who were pushing boundaries of skiing everywhere by pushing them first here. Now, Benny is wearing a ski instructor jacket and working for the resort, and that crew has been changed by death and age and dilution, but he's still breaking the rules when he has a reason to, whether that's terrain choice, or bringing in tagalongs like me. When I think about people who have forged a life around skiing, and who have shaped skiing by doing so, Benny is one of the first people who comes to mind.

"I grew up on vacation," Benny says, and that sounds slightly ridiculous, but it's true. He's from here, and his family helped build Jackson Hole. In 1965, after staring at the faces of the Tetons for more than a decade, developers Alex Morley and Paul McCollister opened Jackson Hole Mountain Resort in what was then economically depressed Teton County, glomming on to the rampant growth of skiing. In 1966, during a decade when hundreds of ski resorts opened, the red box of the tram started shuttling skiers to the top of Rendezvous Mountain. Benny's dad, Colby Wilson, showed up a year later, when they were still selling lots in Teton Village for $10,000. They came with a lifetime ski pass for the owner. Colby started the Hostel X in the heart of the resort village. He'd visited Jackson from Cleveland and fallen in love with the Wyoming ranges. He wanted to build somewhere cheap so that skiers like him could come crash with their families, so he plugged a shoddy boardinghouse into the resort's freak baby mashup of western cabins and faux Bavarian lodges. Benny and the other local kids took ski lessons from the World Cup ski racers and renowned mountaineers who came to work at the mountain, like Olympic gold medalist Pepi Stiegler, who signed on to run the ski school, giving the new resort some cachet.

Jackson Hole was a deliberately American take on ski culture: less fancy, more cowboy than the European fantasy brought over by tan Austrians between the World Wars. Benny was in the thick of it. The rugged Tetons pulled in the most daring skiers of the time, like Bob Smith and Sam Southwick, who were in the early ski movies directed by Dick Barrymore and Warren Miller. They would stay at the Hostel, and teenage Benny would ski with them. In the rock-riddled steeps of Jackson, the filmmakers found that their pubescent tour guide could outski the pros. He would take them into the high-consequence terrain beyond the resort's boundaries. He was out there every day, exploring places no one had skied before, obsessed with untouched turns, and perfectly steep chutes. "I got indoctrinated into that world of do whatever you can so you can ski," Benny says. "Be on a fishing boat for four months in the summer so you can afford to take six off in the winter. Whatever it takes, you'll do." As a teenager he would tell his parents he was going to school, then would walk into Teton Village Sports, the ski shop next door, change into his gear, and disappear up the mountain. As Jackson—and the rest of skiing—expanded, he started exploring out-of-bounds terrain, like the cliffy dark side of Granite Canyon, on the mountain's sheer north flank. He was finding the boundaries of his ability, and of the resort's control.

His only real concern back then was snow conditions. As he got older, and clicked into the machine of the town, he could gin up everything else he needed. He could get a season pass for a six-pack; he lived at the Hostel and worked at restaurants, so he had shelter and food. And he was linked into the underground, trade-based economy that kept the locals viable, so he could barter for anything else. Skill is currency, and talented locals know secrets you can't pay to learn, so often in ski towns the lowest people on the financial ladder have the most clout. You can't show up and buy your way into Granite. That clandestine knowledge, coupled with athletic skill, is what makes someone a real skier, and back then you could basically survive on that. "The bars had to offer free food, so as a bum you could get a meal by migrating. It was all about community and knowing people," Benny says. "My buddies got jobs in ski shops. I was feeding them, they were giving me tunes."

Benny and I spit out the top of the tram, into the granulated wind, and click into our skis. The Rendezvous Bowl opens up wide and pitchy below us and I follow him down into the flat light, trying to keep up with his round, clean, deceptively fast turns. Benny lets the fall line do the work, picking up energy in the apex of every turn. I am essentially straight-lining and barely keeping up. He makes it look effortless, even though I know he's making a million little micro-adjustments.

When we hit the flat, groomed cat track at the bottom of the bowl, he pulls up and stops. "I'm gonna take you somewhere," he says. "But I better not wear my red jacket when I do." He strips off his ski instructor coat, flips it inside out and rezips it with the dark lining on the outside. We skate off along the cat track, picking up speed.

He took me somewhere. I probably shouldn't say where, because part of the trust of the shoulder tap is keeping your trap shut. But after Benny ski-cut the edge of a spiny ridge, to make sure it wouldn't slide, he let me drop first into the creamy throat of a narrow couloir and after the choke point, where I made a couple of nervous, hesitant turns, it opened up into that perfect alchemy of pitch and powder that makes you feel invincible, and like it might always be worth it to chase that feeling.

MAKE IT UP AS WE GO ALONG

When I came west for the first time, I left Boston at the end of October, caravanning with two of my friends from college. I was skittish about commitment, and tied to an unformed idea of adventure that really only existed in my mind. The only thing I was sure of was that I wanted to be outside. I'd pulled in Steph and Katie, because, despite my claims of independence, I was scared to go alone. We headed to the mountains on the promise of a minimum wage job and a free ski pass, which seemed like enough at the time. Across Indiana and Nebraska we'd spot other cars loaded up like ours, headed toward the foothills somewhere beyond the horizon, skis in the racks, trunks packed tight. I switched CD books with Katie somewhere in Ohio, and as the prairie stretched on I belted along to the Talking Heads, taking the words as a sign. "The less we know about it the better, make it up as we go along."

I don't know much about salmon or geese, and what kind of biological force trips up inside of them when the weather changes, sending them upstream or downwind, but I wonder if it's kind of the same for skiers, an intrinsic pull to winter. Sometimes I think there has to be, because there are a million different reasons why people move to the mountains and most of them aren't very good. For the three of us it was a guess, based on a gauzy vision we'd built from ski movies and the buzz we got from the scratchy mountains of New England. Ivan, the raft guide I'd met the summer before, had helped the three of us get jobs scanning lift tickets at Beaver Creek, Vail's younger, more sparkly sister, which sits just west of it along I-70, the mainline to Colorado's mega resorts. He promised that we'd get to ski all the time, and that they'd hire anyone, so we'd be fine. I let that promise shape my romanticized vision of ski bumming. I was obsessed with the idea of exploring somewhere new, and convinced that I would be more myself in the mountains—that place made the person.

Despite the vision in my head, I hadn't been ready for the rib cage of the Rockies, jutting up fast after the flats of Iowa, Nebraska, and eastern Colorado. In the foothills I fought vertigo and rapidly questioned my commitment to a life I didn't quite understand. Altitude vise-gripped my head, and my lips stayed dry no matter how much water I drank.

I'd pictured the tightly-jointed cuteness of the New England ski towns I'd grown up around, the clapboard of Sugarbush or Stowe, but when we drove into the Vail Valley, where the highway is lined by tchotchke shops and sprawling condos, it felt like a schmancy strip mall by comparison. I hadn't realized how constructed this frontier would feel—I didn't really understand then that the idea of the frontier itself was a fantasy. My gut knotted up. This wasn't what it was supposed to look like. I had my first cell phone, but we were otherwise disconnected, unsure of everything, even the location of the grocery store. We went to Walmart to stock up on cereal and toilet paper, and clicked the locks on our employee housing apartment.

It didn't seem like the mountain would be ready to open soon, even though we had to report for ticket scanning training the next day. It all felt artificial. The hill was brown, just a slice of gunned snow running alongside the lifts. On the other side

of the valley, the Gore Range loomed above a strip mall and a
Wendy's, second homes layering the ridges. Later I would learn
that the 10th Mountain Division troops, who came home from
World War II and kick-started the American ski industry, built
the village here on purpose. They oriented the ski hill to the
north to try to hold snow and designed the ski town to mimic
what they'd seen in Europe, tacking Bavarian gingerbread onto
all the buildings, and then giving it an Americanized Wild West
sheen. They constructed this idea of cowboy independence and
commercialized it.

The mountain opened and the days started to flow into one
another. Snow built up, and the dull browns of late fall gave
way to the clear blues of high elevation winter. I began to figure out the flow of the town, and the way the locals moved
through its veins, under the surface. Once I found my own foot-
ing I started to notice the patterns: the older liftie who always
worked the bottom of Chair 6 and kept linty gummy worms
in his pocket for kids; the snowcat's rhythms, which indicated
how soft the snow would be; the locals who knocked out laps
before they had to go to work at cooler, more established jobs
than ticket scanning. They made it look so easy, somehow turning a clunky mess of gear and layers into fluid sartorial coolness.
I, on the other hand, was constantly fumbling with my dated
gear, dropping gloves and poles on the bus, yardsaleing under
the chairlift. I wanted my body to catch up to the way skiing
looked in my brain.

Our jobs were a menial mix of ticket scanning and on-
mountain assistance called "guest services" which meant that
we spent half our days in the lift line trying to explain to unstable tourists what alternating meant. The benefit was that we
had half the day to ski, which I approached with a methodical
precision because being good felt like the key to belonging. I
would bang through bump runs until my knees ached. Katie

was always a better skier than me, and Steph was much stronger, but I had less of a sense of self-preservation, so I threw my body downhill, shin-banged and bruised from eating shit. Days when I didn't have to scan tickets first thing, I'd ski the uncut groomers, making fast, glassy turns, railing from one edge to another, chasing the zero-gravity sink in my stomach that felt something like love.

We settled into a routine that revolved around the weather. Get up early, check the snow, slip on layers of down and polypropylene. Get on the bus to the mountain; get in as many runs as you can. We fell into an evening habit of dollar après and late-night burned quesadillas. We didn't shower much. Our lives pivoted around storms and our priorities were simple. We became skiers like that, sliding into the ruts.

I started to love the glistery gray days when a storm was coming, along with the clear ones that rolled through after. I tracked weather and wind, finding lines through the trees and memorizing specific rollovers where I knew I could catch air. I would tag along with the ski patrollers when they went to open new terrain. The guys, especially the young ones, would usually let us come.

I know now what a stereotype I was. One of the new girls, same every year. Fresh out of college on the East Coast or Midwest, most of whom would head back there after one winter. The ski industry survives on a churn of fresh low-wage workers, who come in for a gap year season to bump chairs or fit rental boots. Long-timers see them come and go, nearly interchangeable, and you have to stick around for a few seasons to cut through the cynicism. But it takes a while to see that as jadedness instead of judginess. The first night, jacked up on attention, I almost went home with a sleepy-eyed snowcat driver from Wisconsin, not yet knowing that newness was a currency I shouldn't squander. Back then all I wanted to be was part of it, capable and inside the system.

Slowly, I wised up to what my priorities should be, following weather patterns instead of people. And slowly—probably in part because of my try-hard ways—I got better. I had been shaped in the stiff upper lip East Coast style, tough enough to stick through ice storms and shitty winters, but when I moved to the mountains and started to operate inside the world of skiing, instead of just on its margins, I let it shape my days and then my years. I told myself I was becoming a skier instead of just someone who skis.

"Trace it far enough and this very moment in your life becomes a rare species, the result of a strange evolution." Rebecca Solnit wrote that about the coincidences that make up our lives, in *The Faraway Nearby*, and when I look back at my own choices I can see the arc of that evolution. When I first drove off I had no idea how strongly skiing would take hold, and how it would shape all my decisions that came after: who to love and where to live and how to present myself in the world.

SWIFT, SILENT, DEEP

In Jackson, if you look closely at the people in the front of the tram line, you'll see that a lot of them wear patches or a pin. A black diamond, a skull with crossed ski poles and the words *Swift. Silent. Deep. 1st Tracks OB.* That insignia comes from Benny, and the people he was skiing with back in the '80s and '90s: the Jackson Hole Air Force. They were the locals who built a legacy of crossing boundaries, and of putting skiing before rules, or real dinner, or a reasonable roof over their heads. What started out as an under-the-radar ragtag gang became a symbol of skiing's obsessive focus, and the camaraderie and chaos that came with it. Keep your eyes on them.

The patches don't mean as much as they used to, but as the Air Force grew, they became synonymous with extreme skiing. Local kids, like Benny and the Hunt brothers, who also grew up in the village, started hanging around with people who had moved to Jackson for the skiing, like Howard Henderson, a med school dropout from Michigan, and Dave "the Wave" Muccino, a tiny guy who went huge. They were the best skiers on the mountain, and they started quietly calling themselves the Jackson Hole Air Force, an offhand joke about how they never kept their skis on the ground.

The Air Forcers skied into what's now called sidecountry, the uncontrolled terrain just outside the resort boundaries. They opened up lines off the back side of the mountain in Granite Canyon, and to the south toward Cody Peak and Rock Springs, mountains ribbed with cliff bands, cut through with chutes, and blown in with snow. As they bore deeper into the terrain, they started to gain a reputation. They were skiing first descents into sustained, near-vertical couloirs, like Corbet's and S&S, which are still big mountain test pieces to this day.

Benny says they were hungry for powder and jacked up on adrenaline. Even as skiing's popularity grew and Jackson's image ascended with it, he says the hard-skiing, hard-partying crew of Air Forcers still held the mountain's secrets. "We were the Hells Angels, we pioneered sidecountry skiing," he says, and that is not an empty, backward-looking brag.

They were ghosts for a long time, unseen, but Jackson is a visual mountain, and every snowfall brings a new chance to make a mark. After a storm, you can see the tracks in the marquee lines where only a couple people have been brave enough to ski. Back then, if you were there enough you could start to guess whose tracks they were. If you paid attention, you'd see those people in the narrow upstairs balcony of Mangy Moose bar, or in the front of the tram line, semisecret but clearly onto something.

You couldn't ask to join, but if you stuck around long enough, and were good enough and kept your eyes open, maybe at some point you'd get a "follow me," and a chance to chase Howie or Dave into the trees. Eventually everyone knew that the Jackson Hole Air Force were the best skiers on the mountain, and maybe in the world. Outside of Jackson they were building a reputation, too. Guys in the Force were winning the early days extreme skiing competitions, starting heli-skiing operations in the Chugach Mountains of Alaska, stretching the limits of the sport.

There were other ski gangs back then. The Dogs of Bell, in Aspen, and the Sunnyside Sliders in Washington State, but the Air Force gained acclaim in part because they attracted the best skiers around—Doug Coombs, often considered the greatest steep skier of all time, was one of the originals recruited by Henderson—and in part because their quest to explore the backcountry became a cops and robbers battle with ski patrol. The Air Force started using radios to eavesdrop on patrol so they could stay ahead, building backcountry huts and smoke shacks right beyond the borders. Taunting the narcs. Skiing in costume. Pointedly aware that they were breaking the rules and getting away with it. Their disregard for authority exemplified the growing rift between the local skiers and the incoming uptight corporate culture of the growing ski industry.

The early Air Forcers knew how to get away with breaking rules while still staying safe, but as their numbers grew, and as extreme skiing became more popular, ski patrol worried less skilled skiers might follow their tracks, setting off avalanches and getting hurt or killed. They had to draw a line somewhere, and the resort didn't want to be responsible for areas they didn't control. It turned into a real fight for a while, when the resort cracked down and started pulling passes. In 1997, Coombs, the most famous extreme skier in the world at the time, got kicked off the mountain for ducking ropes. His banishment set off a wave of negative publicity, which ultimately led the ski area to change its policies. They opened their backcountry boundaries in 1999, which solidified the Air Force's status as pioneers, and also shaped how other resorts manage the margins of their terrain. Now, most ski resorts have some kind of open boundary policy and avalanche control has become a crucial part of patrolling.

How we ski now—hunting powder and steepness, on the edge of control—is thanks, in large part, to the Air Force. Backcountry skiing, out of the bounds of a ski resort, under your own power, is the fastest growing segment in the sport. But to me, the fight with ski patrol, which introduced the world to the Air Force skiers, is the least interesting piece of the Air

Force ski bum lore. I love hearing stories about dropping into the chutes, unsure of what was below or if the cliff bands closed out. I love their off-kilter camaraderie, and the idea that you should be sneaking around, damning the man. But that rebellion can't last forever. After the boundaries were cracked, so was some of the mystique of the Air Force. They were still the best skiers on the mountain, but they had nothing to battle besides inertia and crowds.

HOW IS EVERYONE *NOT* DOING THIS?

In Jackson, greasing the system is often called "Skid Luxury." Skid is Jackson slang for dirtbag, and luxury means living your best life in the mountains—the thing people pin aspirational, expensive vacations to—while still being a cheapskate. Skid Luxury means you're at the front of the line for every storm. It's the epitome of ski bumming.

The credo is often credited to Jeff Leger, a longtime Hostel dweller who came up through the ranks behind Benny. He'll deny credit—although he did create a Skid Luxury logo for a long-ago ski film, but he's been part of the creative backbone of the place since he showed up at the Hostel, unaware of the weird fantasy he was entering, which would suck him in for life.

I meet Leger at the bottom of the Sublette chair. He pulls out his headphones, which are blaring some kind of deep reggae, and smiles huge when I ski down. We ride up, over the near-vertical Alta chutes, then slide out toward Green River, and the resort boundary. Leger is well known, locally, for a lot of reasons. He's the voice on the resort's snow phone—a gig that lets him be at the mountain pretty much every day—and he's also notorious for going bigger and wilder than almost anyone else around. He's still known for throwing huge front flips into the gut of Corbet's Couloir, even though he's in his fifties now, twice as old as most of the other people skiing similarly.

My friend Amy, a photographer who lives here in Jackson, says he has bird bones, so he can launch himself off almost anything and land light. As we drop though the rollers, making quiet turns in the new snow, he skis playfully, popping off little side hits, sparkly and giggly. Leger was an East Coast kid like me. He stumbled into the Hostel when he was eighteen, in the midst of moving west in a winding path from the Boston suburbs. He stayed for fourteen years. He'd head to Bozeman in the summer, to take classes at Montana State and to chase the woman who eventually became his wife, but his life cycled around skiing. He watched Benny and Howard Henderson break the rules—recycling tickets so they could go up the lifts for free, sleeping on the tram dock for first tracks—and he watched them launch down Corbet's or S&S Couloir and he wanted in. "I just glommed on to those guys. As a kid that myth is irresistible," Leger says. "We were playing cops and robbers, but it was like an apprenticeship in serious matters."

He also became the unofficial resident AV geek of the Hostel. He says they'd sit in the dank basement watching VHS tapes on loop—raw footage shot by Benny and company—and it felt like being in the shadow of the world's coolest older brothers. They gave him the twisted keys to the castle, and the idea that somehow—call it destiny or disillusion—he'd ended up in exactly the right place at the right time. "What if the first people I'd met were assholes?" he says. "I could have been selling insurance in Malden right now."

But he's not, and really, watching him ski, I don't think he could be. More than almost anyone I've met in the mountains—maybe even more than Benny—skiing seems elemental to him. There's a purity to his ease and joy. When he talks about carving out a world here, where he can take care of his kid and ski every day, then paint houses in the summer to make it work, it's impossible to imagine him anywhere else. His identity is twined around the mountain.

"What do you like to ski?" he asks, when we stop on the edge of the trees. The air is so thick with humidity that it sparkles, catching the flakes and fragmenting the light. I can tell he's feeling me out, trying not to offend me, but fishing to see if he can take me somewhere interesting. "I'm going here, you go here," he says, pointing, after I try to explain. He gestures over the ridge, and then he's gone, hucking his body over every possible roller and lip, his purple jacket and orange pants a flash of color in the distance.

I ski the slightly mellower chute to his right, and I can hear him whooping as he catches air, bird bones skyborne, knees tucking in the slipstream. We both glide into the bottom of the pitch, and then I'm following him down the rollover of a snow-covered boulder field, trying to keep up. Some kids yell, "Skid luxury!" at Leger as we weave back out onto the rutted traverse track, keeping our knees loose, holding speed.

When we get back to the tram dock, and stake our place in line, Leger tries to explain how he fits into the Jackson Hole geometry, and why it suits him so well. "Skiing encompasses everything for me," he says. "If I need to get jazzed up that helps, if I need to detach that helps, if I need to be feeling at peace it's got that, if I need an adrenaline shot it's got that. There's no time it's not right." He says there's a certain feeling of balance he can't get anywhere else. Maybe you know that feeling, too. Call it harmony or energy or flow, whatever woo-woo explainer you want but it's the sense that everything is aligned, including gravity, when you're driving your skis downhill.

He says learning that feeling was part lucky stumble, part providence. "Even going back to sixth or seventh grade when they started directing you to the guidance counselor, I was calling bullshit on the whole thing," he says. "I'm kind of antisocial, and I don't play by the rules. It's not even rebelliousness, it's that I'm going to figure out my own program. And I don't know why it worked out so well because I certainly didn't have a plan when I came here. All I knew was that this was really interesting. I almost was like, 'How are people *not* doing this?'"

I get that disbelief. I don't live in the everyday, but I know that I still like the clammy adrenaline of the ridge hike, and the aliveness of the inbreath that comes with it. I still crave the daily rhythm of recreation towns. Of being with someone who knows the bartender, of falling a little bit in love with ski patrollers because they're solid and capable. I know those moments are out of reach now, but I still want to feel like every turn might hold adventure.

The ski bum myth is, just that, a myth. It stems from some of the broken stories that are frequently beaten into our brains in US history class—Eurocentric frontier narratives about individual liberty and open spaces, and how the West was "won."

The idea that the western US has ever been open for incoming people to explore and claim is, in itself, deeply problematic, but it's steeped in a pervasive sense of nationalism and American bootstrapping individualism. I've found it creeping everywhere in ski culture, where so many ideas of success or glory are tied to exploring or making a mark on the landscape.

In 1893, historian Frederick Jackson Turner made a speech about the frontier myth, and what he called Westering, shaped by Horace Greeley's credo of "Go west, young man," and of growing up with the country—an expression he borrowed from writer John Babsone Lane Soule. Turner said American identity was directly based in the exploration of the Wild West, and that it was a citizen's duty to go forth, explore, and claim. That's a white, colonialist construct that erases Native American history, and it's thankfully fallen out of favor. But the notion of exploration as a measure of one's ability, one's worthiness, still persists for me, even though it stems from false ideals. I know it's wrong but it's still hard to shake, and it shows up under the surface of how I see myself outside, and why I think it's valuable. On some level I've been indoctrinated into thinking that going to the mountains is important and good.

The myth of untouched exploration tends toward the Pacific. The gold rush and the land rush, and even the startup sparkle of Silicon Valley all stem from the idea of the rangy openness of the West. There's a jingoistic reason why I still say "back East," and I know I'm not the only one.

I have trouble untangling it from the idea of going to the mountains to prove yourself. So many cultures have a coming of age ritual and, especially for men, those rituals often involve pushing yourself physically. Go to the woods, to the wild, to show that you can be a man.

I can relate to Leger's path because I see the ingrained appeal of one thing that can fill almost every need. Part awe, part connection, it's the selfish feeling of wanting to be tied to something wild and beautiful. It feels special and it feels rare.

Especially now, in the age of Google Maps, where every square inch of the planet seems to be documented, and there's nothing left that's unexplored, it feels like we have to create our own romance, and find new ways to tell stories about being bold. Going to the mountains felt like an important tangible test piece of my fortitude and bravery.

Leger's still in the exploration phase, even though he's been here for a long time, but I left when the seasonal stakes got too scary. I didn't want to be fully dependent on my body, especially after a few close calls. The hunt for novelty became harder than it was fun.

But when I drop into Leger's life for a day I get an achy sense of nostalgia for the single-mindedness of my own, younger life. Those feelings are amplified by the stories of untouchable adventurers who came before me. I fetishize the past, even though

I know exploration comes on the back of colonialism and I'm keenly jealous of the generations who got to explore more than I did, before everything was mapped onto devices we keep in our pockets. Leger's temperament and timing made his journey possible, and I wonder if his path is still replicable.

HOSTEL TENDENCIES

"Come over to the Hostel later," Joe Paine says, when I run into him in line for the Bridger Gondola, his red beard crusted white with snow, a maniacal glint in his eye. "We're gonna build an igloo." Joe, who is my age, is the current iteration of Benny and Leger, the guy who had anchored himself in skid luxury. He lived in the Hostel, like the bums before him, for a decade, bartending at night and skiing every day, committed to this place. He's evolving a little—he just moved into his own place down the road in Wilson—but he still stops by the Hostel every morning. He stores his skis in the locker room, skims free coffee, and bullshits with the guys who live there now before his shifts bartending at the Spur, one of the village bars.

I stumble over in the early evening after pilfering free poutine at a ski instructor event Benny snuck me into. I lean my skis up against the worn side of the Hostel, kick snow off my boots, and head into the lobby where it smells like wet wood, old coffee, and sweaty wool. The Hostel skids—the kids who work there in exchange for cheap lodging—are milling around, still half-dressed in ski clothes, cracking PBRs. Joe is sitting in one of the deep chairs, pulling off his boots, captaining the swarm of dudes. Yanni is sort of manning the front desk. The rest of them, like Connor and the two Stevens, are switching layers and making plans.

The Hostel is a not-quite-cohesive blend of local housing and guest lodging, as it has been since Benny's dad built it, more than half a century ago. Tourists bang up and down the stairs, shuttling skis, and carrying some semblance of dinner or half-eaten après. In the basement there's a locker room, a tuning bench, games, a VCR, and the pervasive smell of boot dryers and old saggy couches. Visitors knock around ping-pong balls, and pore over trail maps. But the heart of the social scene is the dudes (mainly dudes) who live here, or have lived here, and whose lives revolve around this place.

Tonight, for reasons that are obscure to me—tradition, or something, Joe says—they're building a giant igloo outside the front door. And, because I'm here, I'm going to help. Hey, why not? We pull our avalanche shovels out of our packs, break into another 30 rack, and head back out into the neon-charged starlight to hack snow and stack blocks.

Joe doesn't live here anymore, but it's still his home base. He's tried to leave Jackson, as have some of the other dudes. One of the Stevens spent a whole winter back in Pittsburgh, another one tried the military, but they kept coming back to Jackson, to the Hostel, because the rhythm and the goals make sense: get up, check snow, go ski. Do work, which might be bone-crunchingly hard, or uninteresting, or happen at ungodly hours, but which lets you ski more. Come back here to a crew that understands that flow. You can live in the future in the off-season, but when it's snowing you just live in the now.

The thing that keeps it burning when you're gone is that you *are* missing out. If you really believe the system is bullshit, like Leger does, and you really think that the Tetons are the most perfect mountains, like Benny, maybe you can make it work without strife. Joe has seen the good and bad parts of the story, and he knows that sometimes it sucks—but he also knows there's something real there.

The Hostel is back behind a row of newer faux barnwood buildings, and the fancy Italian place, with its view of the mountain obscured. The village is a caricature of the American West, pushed through the fondue-choked arteries of pseudo-European lavishness, as if Heidi went elk hunting in Yellowstone. Jackson has been trying to cultivate this image of rustic opulence for more than half a century. In doing so, the town is pricing out the real people. The Hostel is one of the lone tethers to a less-fancy past.

In the narrow pine-lined yard in front of the Hostel we chop blocks out of the snow and stack them into an oblong oval. One of the Stevens decides a wall needs a window seat, so he starts carving out a hole into our new construction. Someone cracks another case. Joe rounds up some more workers, shaking them out from inside. We get ambushed from up above by snowballs and dissolve into giggles. Maybe we're in the low rent district, and maybe none of this is sustainable for long, but for now we're in one of those moments where you wouldn't want to be anywhere else.

After the igloo walls are solid, Joe and I walk over to the Mangy Moose, the old woody bar that's been around almost as long as the resort has, for one more drink. It's late, but when we make our way upstairs to the Penalty Box, the skinny balcony where a lot of the old Air Forcers hang out, Duct Tape Pete and Eric Rohr are still there, talking with Max the bartender. Max nods to Joe and slides a burned pepperoni pizza down the bar, saying that we should eat it so it doesn't go to waste. I'm technically a vegetarian but I wolf it down, because the essence of skid luxury is getting things that other people have to pay for, for free.

Joe tries to trick Max into giving him beers on the house, and I ask Eric and Pete about how they got here, and why they've stayed. Eric, who has been with Teton Gravity Research, a ski film company started by second generation Air Forcers, since almost the beginning, tells stories about what it was like back when they had to sneak out of bounds, and Pete tells me about how his New York jeweler dad reacted when he came west in the '80s. Poorly. When I finally break away from the bar and make my way out to the bus stop it feels like I've been worked through a time warp.

Sitting in the Moose, hearing stories from guys who carved out a little bit of that history before me, I am both transfixed and a little melancholy. I'll never have their life, and I don't think I'd want it. But there is a tenderness to the way that locals take care of each other, even if it's just with cold pizza, and the way that skiing has given them a bond.

I'll see Pete the next day in the tram line and as we pass by each other in the queue, he'll give me the briefest of head nods, a confirmation of being seen, and that we're in this thing together—aren't we the lucky ones.

HOW TO MAKE A MYTH

The history—and story—of skiing

WILL SKI FOR FOOD

My first concrete memory of skiing is the clinic at Mount Snow, Vermont. I'm seven, still in ski boots, gingerly holding my right thumb, broken backward by a fall. I'm not sure if the memory is my own, or if it's one of those family fish tales I've been told, but I remember my father on the chair telling me, ill-advisedly, to stop myself by falling down if I started going too fast. And I remember my mother growling, "JOHN!" when she found us in the clinic, waiting for the plaster of my first cast to harden. Skiing was personally painful and parentally disappointing from the beginning. I still get an achiness in that bone.

My family skied, but in a casual, occasional way, so my obsession had no clear path to follow. I was privy to some of the heavy privilege it takes to become a skier (a family who could afford and enable skiing, and who didn't shun me when I decided to use my college degree to scan lift tickets, for instance) but I wasn't indoctrinated from birth. When I truly threw myself into mountain life, I met people who had been groomed to be skiers since they were babies. They'd been bobble-heading on tiny skis, bashing slalom gates as preteens, and taking week-long ski vacations their whole lives, part of an even loftier level of privilege I'd soon learn was prevalent within the rarefied sport. I met other people whose parents had been ski bums, as had theirs before. I was an East Coast city kid with no real genetic destiny or geographic proximity, but for me the distance may have driven the fetish.

"I am a skier" is a particular kind of self-definition—you don't do it, you are it, both noun and verb. I first claimed it as a teenager. By high school I was already a crush-heavy sucker for talent, but instead of JTT or other pop culture heartthrobs, I fetishized JP Auclair and JF Cusson, the stars of the nascent freeskiing scene, which was starting to crest in the late '90s. Even then I liked feeling like I was onto something special and rare.

I went to a big inner-city high school, known more for its progressive student-parent day care and its liberal politics than for being a contender in elite high-dollar sports, but we had one gym teacher and one science teacher who both loved to ski, and who made it their mission to enable public school skiers. On Saturday mornings in the midwinter Massachusetts dark, Mr. Greenwich and Mr. Haverty would load a bus full of sleepy-eyed teenagers and wake us up as we lumbered into the Sugarbush parking lot. We'd bust out into the nose-hair freezing cold, learning the farthest edges of our speed on the blue ice of Organgrinder, taunting each other from the Castlerock chair. I was constantly trying to convince myself I wasn't scared, already obsessed with speed and skill, chasing the older boys, the better skiers, already pulled by my tangled wanting.

As a scrawny fifteen-year-old I sold breakfast in the cafeteria to raise money for the ski club, so I could ski for free. I took the early morning public humiliation as penance for my desire to get on the ski bus, and be a part of the club, literally and emotionally. Now, we have social media to showcase who we want to be. We brand ourselves easily and implicitly. But back then it wasn't as obvious. You were what you did, or at least I felt like I was. We made club T-shirts that said,

Will Ski For Food, even though our parents were still feeding us, and I wore mine to shreds. I liked the idea of myself as a skier: tough and graceful and talented, all the things I was worried I wasn't. Skiing let me feel like that was valuable, that my toughness was good.

TUCKERMAN TO THE 10TH MOUNTAIN DIVISION

Now, depending on who you ask, skiing is extra bourgeois. It conjures visions of stockbrokers in Oakleys sweatily drinking microbrews on a snowy Vail sundeck, or families from Connecticut vacationing in expensive hotels with massages and room service. But for all the ways it has evolved into excess and become a particular white male conception of power, ski culture, especially in America, is historically tied to the rejection of social norms, and has been since the beginning of last century.

In the US, the ski industry starts where my ski story starts, in the knobby mountains of New England. My first time on backcountry skis, outside the boundaries of a resort, was a teenage attempt at New Hampshire's Tuckerman Ravine, by way of the Sherburne Trail of Mount Washington. The Sherburne winds through brushy forest before crossing a creek and heading up toward a steep bowl. It was one of the first ski runs in the US, hacked out by hand in the 1930s. Back then, runs were cut steep and skinny, just a couple of skis wide. That was skiing for a long time, no lifts, just a grind uphill and a slide back down.

I struggled every step of that first silver cold morning, slipping in the parking lot then sliding even more when the trail got steep. I beat myself hungry and cranky on the icy ribs of the bowl itself, as the sky turned leaden and gray. I was almost too scared and exhausted to take in the wide sweep of the ravine when we stopped for a break at Lunch Rocks. As we skied down, hunting for pockets of powder, the conditions got thwacky and New England tough, but there were enough easy boot-deep turns that by the bottom I was hooked.

Scandinavian immigrants brought Nordic skiing to the US as transportation in the late 1800s. In addition to instigating the trails of New England, they made ski jumping popular at places like Howelsen Hill, in Steamboat Springs, Colorado. In the early days of the 20th century, wealthy Americans traveled to European ski resorts on elite adventures and brought home the idea of skiing as recreation. Alpine skiing started to take root as European skiers from places like Grindelwald, Garmisch, Chamonix, and St. Anton came over to teach Americans to ski.

In 1909, inspired by a winter carnival he saw in Montreal's Laurentians, Fred Harris formed the Dartmouth Outing Club in New Hampshire, which became the initial hub of stateside skiing. Harris was the first person to climb Mount Washington by skis, in 1913, and a year later John Apperson ascended the same ravine I climbed. Those early days are filled with stories of handsome college boys throwing themselves over the Tuckerman headwall, or trekking up mountains that had never been skied before. The idea of first descents in the mountains of New England feels deeply romantic to me. I love the story of the 1939 running of the American Inferno at Tuckerman's, arguably America's first big mountain competition, when Dick Durrance (the best skier on the first US Olympic ski team, who went on to win seventeen national titles and head up operations at Alta and Aspen), faced off against Austrian Toni Matt (the future national champion and coach of the US ski team). While everyone else carved turns down the steeps, Matt aired over the whole face of the headwall to claim a downhill speed record that still stands.

Skiing started to seep into broader culture. In the 1930s, as part of federal efforts to pull the country out of the Depression, the Works Progress Administration and the Civilian Conservation Corps cut ski trails like the Sherburne and the Thunderbolt which plunges down Mount Greylock, the tallest mountain in Massachusetts. As the region became more urbanized, the Appalachian Mountain Club popularized skiing in New England as a form of escapism. Recreation, as it often does, reflected politics. "In the '30s Hitler was on the rise and people wanted to get out of Europe," says historian Annie Gilbert Coleman, who wrote a book about the growth of skiing. Coleman says that Austrian instructor Hannes Schneider, who invented a technique for parallel turns known as the Arlberg technique, fled to the States after getting into trouble with the Nazis. German instructor Otto Schniebs, who coined the credo, "Skiing is not a sport, it's a way of life," started teaching in New England for a similar reason. He was escaping Fascism and looking for a new frontier, and skiing was a way for him to get out.

Skiers like Schneider brought a European influence and technique, but the sport took on an American slant in rangier, wild mountains, just a few years behind the innovation in the Alps. In 1932, Gerhard Muller built the first-ever rope tow in St. Moritz, Switzerland, and the initial American one went up in Woodstock, Vermont, at Suicide Six, in 1934. Ski hills started to pop up all across the region. New England was the center of US skiing from the '30s to the '50s, but as it grew, the sport—like so many other American ideas about adventure—looked west toward the high peaks of the Rockies, Sierra, and Wasatch.

In 1936, Averell Harriman, the executive chairman of Union Pacific Railroad, started Sun Valley Resort, in Ketchum, Idaho, as a ploy to get high-end traffic onto the train. He envisioned it as a fancy celebrity resort—a pocket of that European elegance in the Wild West—and he brought in Steve Hannagan, the PR shark who had previously made Miami Beach famous, to make it so. He hired Austrian ski instructors, who filled the dual role of ambassador and sex symbol, and created a glossy vision of beautiful, fit people on the slopes. It was the first inkling of skiing as an ideal, and its traces of glamour—fancy carpets in the lodge, buffed trails, celebrity sightings—still linger in Sun Valley.

But just as it was starting to get glamorous, World War II halted skiing's cultural ascent. Many of those young men eager to prove themselves in the mountains, including Durrance and Matt (who eventually became an American citizen through his service), went to war. A group of those top skiers ended up in the 10th Mountain Division, an elite force hand-selected for their mountaineering abilities and toughness. In 1941, Charles "Minnie" Dole, who had recently formed the National Ski Patrol, wrote to the War Department and recommended training skiers as soldiers, because they knew how to move in the mountains. In 1942, President Roosevelt approved the idea, and nine months later, Camp Hale, Colorado, a wide valley south of what's now Vail, became the 10th Mountain Division's high mountain base, where they could train at altitude in an unforgiving snowpack.

Despite their serious, expensive training, the skiers were only deployed for a brief period. The soldiers of the 10th went to Italy at the end of 1944, under Major General George P. Hays. In January 1945, they seized German positions in the Apennine Mountains on Riva Ridge and Mount Belvedere. They held on until May, when the Germans surrendered on VE day. But while their battle experience was limited, they had a big impact on skiing.

After the war, the 10th Mountain Division veterans returned to the US obsessed with skiing, and ready to spread it to the States. They'd seen the way skiing tied towns together in Europe—and they wanted it. Over the next few decades, 10th Mountain Division veterans established sixty-two ski schools, and started some of the biggest ski areas in the country. In 1946, Friedl Pfeifer opened Aspen, and built the world's longest chairlift, Lift 1A. Veteran Jack Murphy started Sugarbush, Vermont, in 1958, and Peter Siebert founded Vail in the early 1960s, selling lift tickets that cost five dollars. He named the longest run on the mountain Riva Ridge, after their battles in Italy. The 10th veterans weren't the wealthy elite who had previously dominated the sport. They came from diverse economic backgrounds, and as they opened ski hills and taught skiing they expanded the availability of the sport.

The 10th Mountain veterans democratized skiing, and they also glorified the idea of dropping out to chase an ephemeral, physical life. "After Italy, a lot of them said, 'Fuck it, I'm going to take a job that I like,'" historian Annie Coleman says. "It's like after 9/11 when bankers were moving away from Manhattan." In building the American ski industry, they also created the idea of the American ski bum.

Their dream grew out of a '40s post-war counterculture that rejected stability, and it dovetailed with a growth in national prosperity. Before the war, there weren't a lot of middle-class people who gravitated toward skiing, especially not full-time. It was a hobby for the upper class. As more Americans accrued disposable income they could spend on recreation, ski areas proliferated. Those mountains needed workers, preferably young people willing to perform menial labor in exchange for a ski pass, and those workers became a generation of early ski bums. They gave up stability for adventure and they gained entrance into an elite leisure class through the back door. You could drop out of polite society while dropping into a world where the most valuable currency was skill and nerve. As America broadly embraced traditional family values to try to restabilize after the war, skiing became a way to rebel against that conservatism. It was the same kind of rejection of brutality and convention that led to concurrent culture waves like the Beats, dedicated to a different kind of Zen.

Back then, ski towns were often former mining towns or tiny hamlets that happened to be north-facing and snowy, not much to see. But as those places grew and attracted more visitors, skiers stuck around, and started the restaurants and gear shops that now make up the bulk of mountain-town businesses. The towns changed because of the sport, and tourism became a viable industry. You had to be hardy to be a skier in the '40s—lifts were slow and gear was bad. But a growing economy and a desire for recreation drove rapid growth in the ski industry from the post-war years to the '60s. People had disposable income and vacation time, and there were still mountains that hadn't been mapped or stippled with mansions. Both small ski hills, which made skiing accessible for people other than the rich and famous, and big resorts, like Sun Valley, Aspen, and Vail, grew up together. The rise of skiing also tracked with the baby boom, and the birth of a big generation of potential skiers. As the Forest Service encouraged resort development, to build up economic value in public lands that weren't then being otherwise used, new ski areas popped up.

Skiing exploded from half a million skiers in 1956 to three million over the next decade. During the '60s and the first half of the '70s, the sport grew at a rate of fifteen percent a year. "Everyone who had a business made money, it felt prosperous everywhere," says Seth Masia, the head of the International Skiing History Association. He sees those years as the glory days. The sport was at its most accessible, full of opportunities and exploration. Between 1960 and 1965, 386 new ski resorts sprung up across the country. Growth peaked in the winter of 1963 when big name resorts like Vail, Crested Butte, Park City, and Stratton started spinning chairs. But as an activity turned into an industry, new ski areas envisioned their patrons staying and shopping and partying, not just skiing. When Bill Janss opened Snowmass, the last of the four Aspen resorts, in 1967, it evolved around a condo-centric base area. Ski towns, especially the big ones, changed from under-the-radar hamlets to high-dollar destinations. Second homes and rental real estate became a major planning play, and corporate ownership, which didn't necessarily value local town culture or residents, drove expansion.

Skiing's growth started to slow down in the mid-1970s, due to broader economic changes and a tapering population. "OPEC hit, inflation went out of control, and skiing plateaued," Masia says. "And it hasn't grown much, if at all, ever since. Maybe one percent a year." US skier visits have been in the 50-million-a-year range since the National Ski Area Association started counting in 1978, and the average skier age has increased. It was never as popular, cool, and young as it was in the '70s.

EAT THE RICH

Skiing encompasses so many things, from downhill racing to mountaineering descents, but the local lifestyle of ski towns is what captured the zeitgeist. *SKI* magazine first covered ski bums in 1948, and *Life* did a ski bum story in 1950, bringing the image into the mainstream. Leisure was increasingly available and cool. In tandem with the rise of the surf bum, the ski bum became a character in a broader narrative about young people living in mountain chalets and skiing every day. Cleaning rooms, sure, but also drinking cocktails with the celebrities that came through town. Gay Talese covered the high-rolling party scene at Sugarbush, which was nicknamed Mascara Mountain, because the Hollywood hotshots who visited always had their faces on. The Kennedys started coming to Aspen. Dr. Ruth met her husband on a T-bar in the Catskills when they were paired together because they were both short.

Ski resorts grappled with a conflicting tension to feel both adventurous and safe, a dichotomy that still exists today. Resorts try to preserve some of their mountain ruggedness, but they also have to manage risk. Ski mountains are carefully manicured to feel wild.

★ ★ ★

When Dick Durrance, of Tuckerman's American Inferno fame, cut Ruthie's Run on Aspen's Ajax Mountain in 1949, he had the steep old trails of his New England glory days in mind, but he also wanted to make the terrain friendly for visitors, with views of town, because that ease brought in money.

As the original ski bum generations—boomers and their parents—started to age, the drop-off in growth changed how resorts were run, because suddenly those resorts were squeezing money out of a static population. In 1976, Harry Bass bought Vail and pushed out Pete Siebert, the 10th Mountain Division member who had founded it. As Siebert's generation aged out or moved on, mountain management largely shifted from 10th Mountain veterans, to real estate developers with MBAs. Because of the thin margins on lift tickets, the fantasy shifted to selling vacations instead of simply building ski hills. Resorts began to consolidate. Ralston Purina bought Keystone and Arapahoe Basin, in Summit County, Colorado, in 1973, and 20th Century Fox bought Aspen Skiing Company in 1978. Transportation improved, making weekend trips easier. By 1982, when Beaver Creek and Deer Valley—the last big destination resorts—were built, they were fully designed to cater to high-dollar, out-of-town visitors who wanted heated walkways and fancy seafood restaurants. Skiing became synonymous with luxury. That's why Jackson today looks like a bunch of cowboys raided the bank at Goldman Sachs.

It became more litigious, too. In 1974, James Sunday sued Stratton Mountain, when he tripped over a branch on the trail and became paralyzed. His suit set a legal precedent that still stands. Ski areas became liable for accidents that happened within their boundaries, which made every facet of management extra cautious and more expensive. Now you sign a waiver every time you buy a ticket, even though the majority of resorts operate on public land.

As the industry became more corporate and careful, the ski bum idea evolved, too. Towns stratified and self-sorted. You had the fancy people on vacation, and the folks working service jobs, trying to make it work. The gulf between them widened, and it became harder and more precarious to become a fervent local skier. "The first time I saw an article about the death of the ski bum was probably the 1970s when real estate became expensive," Masia says. "It became more and more difficult to live in ski towns if all you wanted to do was ski. Instructors and ski patrollers started to get pushed out." The under-the-radar ski bum took on a more symbolic weight. It came to represent the truehearted soul skier, the opposite of the vacationing hack.

As it became more difficult, the romance surrounding the ski bum grew. Ernest Hemingway, who lived in Sun Valley, wrote about avalanche danger, even sending his signature short story character, Nick Adams, out into the backcountry. James Salter wrote "A Homage to a Legend Called Aspen" while he lived there. He'd have dinner and talk writing with Hunter S. Thompson, who also lived in town. Thompson, reportedly, only skied once, but he was a professional partier, and a figure in the Aspen après scene. Their stories, along with ones by Cheever, Updike, and others, crystalized the ski town fantasy in fiction. They made grist of the skid, and his rejection of capitalistic tendencies. The ski bum became a literary hero, smearing perception with reality.

STORYTELLERS

That hero narrative has colored the broad idea of what skiing can be. Almost everyone I talked to, regardless of their age, said they were the last generation to really live the dream, and that the history, and the highwire feats that came before, are more interesting than the present. In Jackson Hole, a second-gen Air Forcer, filmmaker Jon "JK" Klaczkiewicz, understands the fairy-tale quality of the sport. "The legend grew and grew, you learned from being in the lineup every day, and part of it almost didn't even seem real," he says.

JK came to Wyoming after meeting some of the original Force while he was interning at *FREEZE* magazine in Boulder, Colorado, in the late '90s. He was debating between moving to Jackson or Squaw Valley after college, and the whispers of a secret society of the best skiers in the world pulled him to Wyoming.

You couldn't ask your way into the Air Force, and you definitely couldn't brag your way in. You had to let your skiing speak for itself. JK says you had to prove that you could ski, and that you could listen, and you could hang. When Benny handed him the pin, he felt like he had broken into an untouchable place. "Those early days are hard to put into words, it was such an honor to be accepted," he says. "I remember the first time I went out, it felt like I was able to access a magical world. They told you how to do it and how to sneak back in. It was a heightened level of experience." He thinks that's why the arc of the Air Force is so compelling, it's Robin Hood on snow.

Nine years after the resort opened the backcountry gates, JK made a film about the Air Force and their legacy of rule breaking. In *Swift. Silent. Deep.*, he tracked Benny's early obsession with Pepi, explained the cops and robbers fights with patrol, and followed the cinematic arc of Coombs's career. He pulled the pieces together into a classic quest story. The underdog eventually got the win.

He wanted to freeze that moment because it felt like a distillation of the sport, bigger than Jackson. "It's the quintessential ski bum story to me," he says. "It was early-morning tram lines, and bell-to-bell, top-to-bottom fast laps before sponsored skiers were a thing. It was about doing it for yourself, leaving your signature on the mountain."

He says he got some shit for telling the story to the public because he cracked the legend a bit by talking about it. That's the problem. You have to ruin the secrets to talk about them. But he says the negative feedback didn't come from the original Air Force guys, like Benny. It came from younger, newly-transplanted locals, who were trying to carve out their own space in Jackson Hole, and still wanted to feel like they were in on something secret and special.

Warren Miller, who both lived the dirtbag ideal and commodified it, is credited with creating the image of skiing as freedom. Miller's story is an amalgam of the bootstrapping American dream and the dirtbag ski fantasy. In 1937, as a Boy Scout from a broken family, he took a few turns at California's Mt. Waterman on gangly wooden skis and fell in love. By 1946, after a stint in the Navy, he and his best friend Ward Baker set off on a skiing road trip, towing a teardrop trailer behind an old Buick

Phaeton. After short stints at Yosemite and Alta, they ended up in the Sun Valley parking lot, on a tip-off from a few girls they met in Utah. They sweet-talked Pappy the mountain manager into letting them camp there for the season. In his autobiography, Miller says that Pappy let them stay because he thought they'd be good entertainment. He and Baker ate rabbits and ducks they'd shot the fall before and kept in cold storage under the trailer. They existed on earnings from Baker's part-time job and Miller's amateur artwork, often eating what Miller called Ms. Nicelunchowski: a mix of ketchup, water, and oyster crackers. In Averell Harriman's Sun Valley fantasyland they spent the winter bilking rich starlets and society queens into taking them to dinner, or at least sneaking them into the heated pool. Miller taught skiing to people like *High Noon* star Gary Cooper, and he started filming the skiers. Over the next few years he pulled together enough footage to make a movie.

In 1950, Miller convinced a theater in Pasadena, California, to show the film he dubbed *Deep and Light* and let him narrate it as it played. "I walked out on the stage to face the crowd and started my introduction—a tale of the ski bum life: living in a small trailer in the Sun Valley parking lot, spending only $18 total for lift tickets for four months' skiing," Miller said about the premiere. "I talked about the rotary plow that threw our buried rabbit skins into the tree above the trailer and the intricate dance routine involved in getting undressed outside in the freezing night air in order to get into bed in the freezing cold teardrop trailer after a dinner date. The audience laughed at my stories, not just polite laughs, but amazingly loud laughter."

He'd found a way for outsiders to consume mountain landscapes, and to tell the shiny parts of the ski bum story. The romance was so strong that he turned it into a sixty-year film career, indoctrinating generations of skiers, and convincing them to move to the mountains in the process. "If you don't do it now, you'll be another year older when you do," he taunted in every film. It has since become one of the most enduring messages of skiing. Forget the vacationers, or the over-trained racers, if you were a dirtbag getting by like Miller you were doing it right. *This* was skiing, the thing the 10th Mountain Division soldiers scrapped for, and generations of skiers after tried to achieve. Miller spooled it together in a tidy ninety-minute narrative.

Every fall, right when skiers started anticipating the first real storms of the season, Miller would release a new film: *Steep and Deep, Endless Winter, Beyond the Edge*. Each one was slightly different—the destinations and the skiers changed—but the rhythm remained the same: an aspirational far-off locale, color commentary on that year's hotshot athletes and some light teasing of gapers, the kooks from the city. Every year, still, thousands of skiers sitting in theaters around the country take in that message, just as the snow is starting to stick.

Miller often gets called the original ski bum, but bumming was already in motion in a lot of nascent ski towns before he showed up. When he and Baker moved into their trailer, they'd already passed through the powder skiing scene at Alta, and a handful of the small resorts that were starting to crop up in the Sierra. And, according to Chris "CP" Patterson, the current director of Warren Miller Entertainment, Miller hated that title. He says he wasn't a bum. Even when he was living on the margins, he was hustling.

"He was sort of defensive," CP says. "It was an era thing, but he said, 'A bum to me is a guy who doesn't work, someone who doesn't have any ambition. When I was living in the parking lot, I wrote three books, I was selling photos, starting my film career, and teaching skiing. I had five jobs, I wasn't a bum.'" Even back then the ski bum identity was performative. Tape was cut and carefully narrated to show the glory of being a skier, and poke fun at the gapers who didn't get it.

CP says the hotel managers in Sun Valley understood that cachet of the insider, and the ways the seemingly-carefree locals, with their secrets and skills, enforced the romance: everyone wanted to be with them or be like them. Miller's brilliance was in winkingly condensing the story. The first films were flawless in their silent simplicity: perfect turns, beautiful people, beautiful places.

CP knows the pull of those stories because he knows the exact moment they hooked him. He was six, on family vacation in Winter Park, Colorado, where the hotel would show old Warren Miller movies every night. He'd sneak into the theater alone each evening, pulled in by the cinematic gloss of the ski world. After college he moved to Steamboat Springs, Colorado, where he worked as a projectionist for the film tour, and he's been making ski movies ever since, sustaining the idea.

Miller created a model that persists. Ski season still starts when the ski movies hit, both from Warren Miller Entertainment, and the slew of other production companies that followed, started by people like JK and the Jones Brothers in Jackson, who formed Teton Gravity Research, or the Matchstick Productions crew in Crested Butte. The films themselves can feel formulaic, but they set a seasonal pattern. They get skiers into the same room, listening to stories, just as the air is cooling off and the antsy stretch of fall is coming. And retelling stories is exactly how ski bums are molded and made. Those yearly films became touchstones for people like me. I grew up on their images of powdery pillow drops and steep couloirs, the dash of carnage, the token characters, the suggestion that dropping into sheer Alaskan spines might be within reach. The movies made it look like every day on skis was perfect.

The storylines evolved with the sport. I grew up in the X-Games dawn of freeskiing, just as park skiing was becoming cool, so I remember the yellow-and-black flash of the Solomon 1080, the first twin-tipped ski, and the rise of street skiing, where skiers inspired by skate culture started hitting urban jumps and rails. In the '70s, the films were full of sexy straight skiers who spread the idea that nothing was better than ripping downhill straight into après. The neon hot-doggers of the '80s made it seem like the mountains were always sunny, and there were always cliffs to jump off of. In the '90s, Doug Coombs and his friends were skiing impossible looking lines for the first time, and drinking it blue in the scrungy bars of soggy Chugach towns, waiting to ski those big Alaskan faces. "Warren Miller is the man who made the snowball that created the whole industry," Dirk Collins, a second generation Air Forcer told *Outside* magazine in 2004. Miller made it seem like everyone should be angling toward living in a parking lot trailer, skiing every day.

THE TRUTH BEHIND
TJ AND DEX

The movies glossed over the gritty parts, and the underlayer of inequality in skiing. But together they chart the evolution of an image. If you skim through them you can see the outfits get brighter and tighter, then baggy and loose, you can watch the slope angles steepen, and the featured skiers change from hotshot locals to sponsored athletes. The movies are always just a step behind the real ski town culture. "The '70s might have been the peak. It's nostalgia through a fuzzy lens, but I do think it was a legitimate golden period. Housing was cheap, and jobs were plentiful," says Greg DiTrinco, the former editor of *SKI* magazine (and my old boss) who moved back and forth between Sun Valley and Aspen starting in the late '70s. In his twenties, in between gigs working at the Aspen newspaper, he edited for Hunter S. Thompson, renowned loose-cannon gonzo journalist, which was an adventure all its own. While Thompson ran for sheriff on the "Freak Power" ticket, galvanizing the rebels who moved to the mountains to get loose, DiTrinco saw the clash between ski bums, hippies, burnouts, party boys, and cowboys: the factors that shaped the wild Aspen scene of the '70s. He thinks pretty much everyone will harken back to their twenties as the best period for being a ski bum, but he says being in Aspen back then was a wild blend of looseness, adrenaline, and desire. "All the stuff you saw in the movies is true," he says. "You'd go to the J bar, and there were people doing lines on the bar. That's how you got a table, that was currency. There were older women just strolling in, and I never got hit on more in my life."

It wasn't just in Aspen, where competing ski gangs ruled the mountain and the party scene. After Vietnam, a generation of young men came back from war and a cadre of them beelined for the mountains, ready to remove themselves from society, like the 10th Mountain veterans had done. They went to Telluride, Tahoe, and the Tetons.

The '80s ushered in the Reagan years of rich white dudes and yuppie nose drugs. The goofy sex appeal of mainstream ski movies like *Ski School*, *Ski Patrol*, and *Hot Dog* fed into the oversimplified fallacy that ski town life was all sunny backscratchers, babes in one-pieces, and beating the bad guy to the finish line. Ski vacations became more expensive and elite. Ski towns solidified into resorts.

You can see so much of that in *Aspen Extreme*. The director, Patrick Hasburgh, knew the outline of the story, because the plot tracks his own life. The film follows TJ Burke and Dexter Rutecki, two burned-out Michigan auto workers who move to Aspen on a *Powder* magazine–fed whim. They're immediately sorted into the hot-or-not stratification of ski town society. TJ gets to play with the rich kids on the strength of his cheekbones and ski skills; Dex gets shunted to the underbelly of children's ski school and sucked into cocaine nights. They deal with inequality and consequence, and with feeling like they can never have enough. The movie was panned when it was released for being loose and sloppy, and sure it's long, and the plot swirls, but in watching it now, nearly three decades after it came out, it still feels shockingly relevant. It's about risk and relationships and rich kids, but Hasburgh pulls in the struggle to find housing, the partying, and the proving—facets of ski town life that still resonate. There are romantic subplots, and insider/outsider tension, but the most significant relationship is the friendship between TJ and Dex, and the ways it snaps and changes over time.

I came to skiing slightly after that, so to me, the freeskiing rebellion of the late '90s and early 2000s seems like glory days, because I could hear the firsthand stories of how good it was just before I got there. I get the '70s sheen and the '80s party through photos and videos, but at the turn of the millennium I was there, trying to catch a little bit of air on my first pair of shaped skis, absorbing the magazines and watching the VHS tapes of TGR films in the hot late days of summer, when I couldn't stand to be so far from winter anymore. Until embarrassingly recently, there were pages of old *FREEZE* magazines tacked to one of the pale pink walls in my childhood bedroom at my parents' house. There's a photo of Evan Raps in a red jacket on a pair of Rossignols, airing out above the Stowe Halfpipe. Paper evidence of a preteen crush. That millennial period was another pivot for the sport. Snowboarding became mainstream, and park skiing followed, drawing creativity from skate and surf culture. I watched Mike Douglas and the New Canadian Air Force fighting physics in the park, and big mountain skiers like Shane McConkey and Seth Morrison skiing naked, hucking themselves off cliffs. Skiing felt full of life and in my adolescent way I inhaled as much of it as I could.

Since then, freeskiing has become an Olympic sport, a progressive one even, when American silver medalist Gus Kenworthy kissed his boyfriend after his slopestyle run in 2015, becoming one of the first visible gay male action sports athletes. Technology evolved, too. Cameras got better and cheaper, and GoPros became omnipresent, so everyone could make their own edit and shape their own narrative. I'm curious what the future will look like, and how much of the historical precedent can, and should, hold. We're starting to confront the problematic past. In 2020, Squaw Valley announced it was changing its name due to the historically offensive nature of the word *squaw* and working with local Indigenous communities who first settled this valley to find a more suitable new name. It feels like skiing, finally, is trying to keep up with the times.

History hits the present for me in Jackson when ski movie star Rachael Burks lets me sleep in her second bedroom. I knew of Burks before I knew her. She was the person I idolized on film, brave and cool and fun, just as good as the guys. Burks goes big—even her laugh is huge—and she broke into ski films by throwing laid-out backflips when few other women were doing them. There is harrowing footage of her getting buried under snow for more than ten minutes after hucking a fifty-foot cliff in the Grand Targhee backcountry. But she's gentle, too, she makes me breakfast sandwiches every morning and teases me for not waxing my skis enough, and when she doesn't have to work—she's juggling restaurant shifts, athlete responsibilities, and studying for a real estate exam—she skis with me.

The mass of Jackson's ribby, steep in-bounds slopes predominantly face southeast, staring out at the hollow floodplain of the Snake River, and the undulation of the Gros Ventres range across the valley. When it snows, it's dreamy skiing, but when it doesn't, it becomes the world's steepest coral reef, sun crackled, and chicken-headed by light and heat. It will never go easy on you. The first morning, when we head off the top of the Thunder chair, I boff myself on a bump, yard sale, bury a pole, and stuff my goggles full of enough snow that I beg Burks to go on without me. She tells me to shut up, waits patiently, and then I follow her out toward Rock Springs where she finds us pockets of hot powder. I can hear her huge cackling giggle from downhill as she flies through the deep spots.

It feels a little unreal, skiing with someone I have idolized from afar. Even though Burks is arguably one of the best skiers alive, it's never been easy for her to make it as an athlete—in large part because she's a woman and she hasn't been comfortable playing to the tropes that ski movies have historically laid out for us. Those shiny stories we see in movies made it seem much easier than it is.

It's worth noting that a movie called *The Last of the Ski Bums* came out in 1969. Skiing has always been about pinning down the ephemeral. That night, while Burks makes dinner, she says she's not sure how the future is going to shake out, and why it feels so good in the moment, yet tricky in the long term. "But man," she says, eyes glinting. "Wasn't today fun?"

THIS IS YOUR BRAIN ON SKIING

Who wants to be a ski bum

APRÈSPUNCTURE

Marcus Fuller tells me that Jimi Hendrix came to him in a dream and told him to take up the tuba, and the thing is, I believe him. We are sitting in a warehouse in Bozeman, Montana, next to a pile of instruments he found at a yard sale, still in our ski gear, getting what Marcus calls aprèspuncture. Which means that I just let a man who I've known for three days—and who is currently explaining his Hendrix-spawned brass fantasy—stick an acupuncture needle into my forehead.

I met Marcus on the ridge at Bridger Bowl Ski Area, during a last lap hike at the end of the day of skiing off the Schlasman's lift. We trudged up the boot pack toward the ridge as low clouds flattened the light, turning the valley dusky gold. Bridger traces the skinny spine of its namesake range, and the ridge is convoluted and filled with closeouts and chutes. From the top it feels like all of Montana stretches out in front of you: passes and valleys, a sweep of mountains: the Gallatins, the Crazies, and the Tobacco Roots.

Schlasman's stops spinning at 2:30, and from there, the obsessive locals hike the ridge above it for one more lap before the rest of the mountain closes for the day. I clicked into the daily ritual of a late-day run through a zone called D-Route because

I'd been following a guy named Bubba I met at the bottom of the mountain my first morning at Bridger. Bubba is a transplanted Vermonter whose nickname belies both his skinniness and speed (he was overweight as a kid, he tells me). He farms in the summer, and works events for the ski area in the winter to get a free ski pass. Other than that, he mainly just skis bell to bell, chasing the older folks that he says are the real ski bums. His winter is narrow, and corner-cut, and he avoids a lot of standard markers or stability—hitchhiking up the road to the resort, so he doesn't have to deal with a car, recycling used ski gear, foregoing a steady paycheck—so he can be up here every day. But when I tell him about my mission to try to track down ski bums Bubba says that despite his dedication, he doesn't think he qualifies as a real skid. He's too green, but he knows some people I might want to meet.

We head toward Schlasman's, the lift that accesses the steep, sometimes sketchy terrain along the southern boundary of the resort. After a few runs, Bubba ditches me to go hike toward Saddle Peak, the out-of-bounds summit farther south along the ridge, but before he takes off, he introduces me to some of the scrappy locals who are here every day. Zen Master Marcus grew up here, and has made a mission out of staying, even when money or snowpack might be thin. The ancient ridge hippies, Pammy and Turk Comstock, have been beating down turns in the Bridgers since forever. They're here every day, hewn to the rhythms of the season.

Skiing is the thing that makes the most sense to them, and they do it like drinking water, effortless and necessary. They have an ingrained knowledge of the mountain, of where the stashes are and where the gawky visitors aren't. So at the end of the day when they all convene at the top of the chair and ask, "Hey, want to go for a walk?" I am both smart and dumb enough to say yes. I put my skis on my shoulder and get into the boot pack behind them.

We stop for a breather on the ridge, at a point they call Fort Benton, looking west toward Manhattan, Montana, somewhere in the distance. Bubba, in his sweaty flannel, has joined back up. I'm in my too-warm Gore-Tex, unzipping all my vents. Someone cracks a beer and passes it around. Marcus has a long reddish braid and a bit of a beard that blows in the updraft. It is 3:30? Tuesday? Are those details even relevant? Maybe we're missing out on other things in our single-minded focus on skiing—but right now there's nowhere that sounds better than here.

I can feel the hum of devotion, but I know that it takes a lot of life geometry to get there. Who sustains this life over years of skinny winters and broken bones and sporadic paychecks? How do you get to be Marcus? Today feels purposeful and charged, like we are on a mission together, but that's not always the case. I want to know why the charge wears out for some people, but not for others, and if it's focus or negligence or some genetic obsession that makes it so.

We head into the steep runneled chutes that Bridger is known for, which are sketchier now than they often are because the snowpack is low and unforgiving. While I was in Jackson skiing heavy wet storms, Montana had been high and dry, awfully warm for January, gapped over by La Niña. I follow them to the top of a gully, skiing thin snow over skiddy shale, feeling the rocks ripping up the bases of my skis, brimming with nerves. I am trying to gauge how much speed to hold when I don't know what's beneath me.

Travis Andersen, a local photographer who moved out from South Dakota in 1990 and never left, tells me that you have to be facing to the right when you drop in, because there will be a place where you need to point your skis to clear a gap over a cliffed-out gully. You really don't want to be backward there. Don't look down, he tells me, and hold your speed. Then he drops out of sight and there's nothing I can do but follow.

I look down. But I am going fast enough that I couldn't have wavered if I'd wanted to, and by the time I realize what had happened I am down in the flats, gunning for the base area. I follow as they point it straight to the Griz, the loveable, locally-owned ramshackle bar at the bottom of the lift, where we get pitchers and pizza. Then we go back to Travis's workshop in town, because the other guys have goaded Marcus (who at some point went to acupuncture school, they tell me), into breaking out his acupuncture kit, and giving us a dose. Marcus puts on his white coat and turns on some music that he says will tune into our brain vibrations.

I pull off my ski sock and hike up my pant leg so Marcus can stick a needle into my shin, and he tells me he had a vision about me. I knew you were coming, he says. He saw me hiking to high places, he says. Taos, Alta, the Tetons, to connect with people. And maybe I'm too many strong Montana beers deep, or maybe the needles are doing something to both of our brains, but I suck in my breath. Because maybe he does know something. What else is this but a vision quest?

In Bozeman I'm staying with friends who work as ski guides, and who have spent enough past nights on my living room floor between their seasonal jobs that they're ready to repay the favor. In the morning they make me breakfast tacos before I head up the curve of the canyon to the ski area, where the ridge juts up on your left, vertical and sharp in contrast to the buffed-out hills around it. Bridger is a bit of an anomaly, both geologically and as a business. The skiing is high consequence and hard, and the mountain, with its narrow close-out chutes, has cut the teeth of some of the best skiers of all time, like big mountain pioneers Doug Coombs and Scot Schmidt. The ski area is a nonprofit, run by a volunteer association, which means that, despite the fact that Bozeman is in the midst of a population boom, and other resorts, like nearby Big Sky, have leaned into tourist-based resort activities, Bridger is still underdeveloped, rootsy, and relatively cheap. I'd come up across the Tetons to see if the idea I saw crystalized in Jackson holds here, too.

Marcus was right, I was searching, trying to make some sense of that psychological itch that still sometimes drives me crazy. I gave up on ski bum life because I was anxious about the path it was putting me on. I was stressed out by shifty seasonal income, scared of wrecking my body and worried that my life would only hit one note. But skiing is still woven into the weird tapestry of my life. It threads through all my relationships, my job, and the way I spend my time. I haven't really let it go, and that constant tension made me wonder if there was something in my, or really anyone's, nature that might make them particularly well suited to ski bumming.

I know that skiing is ephemeral and selfish, but I ache when I'm away from it for too long, and I don't think it's just the dopamine drop that drives the fixation. When I'd packed up the car and planned out this loop through the mountains, the question took me farther than I thought it might. I didn't have a date to go home, and I was excited to feel that old buzz of adventure, but on the road solo, as the windshield wipers thrashed past the turnoffs to Deer Lodge or Driggs, the question in my mind morphed from *if* skiing was important to *why* and what I might be missing. These days I am sedentary, antsy, halfway to seventy, staring at a screen too much, and still not quite sure of what I want to be when I grow up, or where. I have questions that my own life hasn't answered and I'm still wondering if skiing can.

Once, in the middle of a chest-deep powder morning, someone asked me what my best ski day was and I said today, of course, but in truth I have a reel of best days dating back to my preteen years. The obsession didn't come upon me in one bright flash. It was a series of moments that spiraled around me until they accrued shell-like. I cracked my ski bum husk when I moved out of the mountains, but I can still feel the pieces, and sometimes, when I slide onto the lift next to someone like Bubba, those pieces reassemble. I get a sense of certainty and identity from skiing. My brain slows down when my body has more space. I like the kinetic energy of driving up a canyon, waiting to see the mountain, anticipation embedded in the landscape. Maybe it's contrast, the balance between surface and sky that makes my ribs get tight, like there is a line pulling taut between a place and me.

ANTISOCIAL BEHAVIOR

Bridger has always been a proving ground. When Schlasman's lift opened in 2008, it made access to the south end of the ridge easier. It softened things up a little. But the north end of the ridge, which was the original domain of the ridge hippies who would launch down the narrow, chalky chutes, is still earn-your-turns only. To get there, you ride to the top of the Bridger chair, strap your skis on your back or throw them over your shoulder, and start hiking up a steep, nearly vertical boot pack to the ridge.

I don't want to neglect history, so on my second day at Bridger, I make a pilgrimage to the north end of the ridge. It hasn't snowed in a while, and the boot pack is impenetrable and slippery. Every step requires a careful, climber-like balance. I am immediately sweating and trying not to tip backward into the void. At the top I meet up with Graham Turnage, a patroller friend of a friend, who said he'd show me around. He waits patiently for me to catch my breath before we head north, the ridge spilling out all around us.

Partway out, when the path gets narrow, another patroller, Shane Cottom, catches up to us, and slows down to chat. Graham says Shane knows the ridge better than almost everyone, and so, to try to buy myself some breathing time, I prod him for his secrets. He's from a family of potato farmers, and grew up east of here on the plains, he says. He tried college for a bit, but it didn't stick, and so he ended up here, where the day-to-day makes sense to him. He deflects my other questions and I take his modesty as the sign of a true local.

I follow Graham and Shane to the north boundary, and Shane keeps going out past the ropes. Once he's out of earshot, Graham tells me that Shane is the best skier on the mountain, even though he'd never say that himself. Shane's going out alone to check on some lines—stuff no one really skis because it's so technical and committing—to see if they'll be skiable this season. He says Shane is always on a semi-obsessive, self-generated mission, which keeps driving him even though he can't quite explain it.

Graham and I ski Hidden Gully, a north-facing chute with a long fall line. We had peered over the cornice at the top, but since it hadn't snowed in a while, the main entrance was a luge, less than a ski-width wide. Maybe if you know this zone you can point your skis through the choke of the chute and be confident that it opens up at the bottom, but I don't and I can't, so we ski around to a lower slot, and traverse into the gully. The entrance is steep. I slide and fumble my way in, cautious on my edges, but once the chute opens up, I can link turns together, and by the bottom the snow is chalky and smooth. A big part of skiing is commitment, in the time spent getting to know the mountain and the decision to throw yourself in.

WHAT COULD BE BETTER THAN THIS?

Commitment comes over time. My first year working menial labor on the mountain I was part of a wave of rookies. We were all bright new gapers, slightly out of place. A lot of those people left in the spring, back to Minneapolis or St. Louis or Maine, for real jobs and relationships. In the mountain town hierarchy, the single season kids are the worst. They're just above the fumbling, entitled tourists. But if you stick around for a summer, then another winter, people start to treat you differently. You do it, then you keep doing it, and your life takes on a certain arc—you're a skier for real.

By my second season I'd developed muscle memory in the clicking of my boots, and a quick skate through the lift line. A new wave of first year lifties and ticket scanners rolled in, and I eyed them with spiky territorialism. I didn't want to be associated with the new girls, I wanted to draw a line around myself, to prove that I was committed, that I knew the ropes. I have always been a know-it-all but I'm sure I was particularly insufferable that winter.

I beat myself blue on the bumps of Grouse Mountain, trying to teach myself to look effortlessness. That should have been a sign that I've always been too much of a try-hard to really make it work, but after hours of hacking turns I had a stake. This was *my* mountain now. I had favorite runs, I had sneaky hideouts, I had a much older ski patroller boyfriend, who made me feel legitimate, even though our relationship underlined every stereotype. Even as he helped me hone my backcountry skills, it was hard to shake the new-girl tagalong feeling. I was constantly insecure about my rank and place. But I had a second job at a pizza place with a contingent of local barflies, who started to remember my name. In a small town with a highly transient population the people who stick around cleave tight. I started to recognize the terrain park crew when they came in for happy hour. I begged the old patrollers for stories, and learned which hotel pools you could sneak into. There was novelty in the rhythm. But often I wasn't sure if I was happy or I was just telling myself I was happy, or if I was just trying to feel anything at all. Was pinging from one adrenaline rush to the next enough to pull me through?

At the end of the day, after the lifts close, Graham takes me into the locker room where the patrollers are drinking PBRs while they take off their boots, the stickered metal doors of their lockers clanging shut. It's a scene I know well. I started patrolling once I worked my way out of ticket scanning and guest services. It felt more legitimate, more purposeful—a step up in the social structure of the mountain. I wanted to belong. But even that desire wore thin after a while. I learned the skills and motions, but I hated it when people got hurt, and I almost never felt fully engaged, even though I still try to hold a claim on that community.

At some ski areas, the patrol is young and transient because there's not much to keep them there aside from the ego boost. But at Bridger plenty of patrollers have been there since the '70s. They built the snow safety program here and across the country, and they have a geographic knowledge that comes from years of controlling and expanding the terrain along the ridge. Once I crack my beer and work up my nerve to ask why they've stayed for so long, someone tells me that they've found a way through the system to escape the system—they've figured out a way to get paid to ski every day—and what could be better than that.

What could be better than that? Sitting here, legs a little sore after a big, beautiful day, I say not much. But I know that's a tricky question. Some of the guys who have been doing it for decades have support networks: houses they bought before Bozeman blew up, supportive spouses, a cushion of family money. Bridger is special because it has stayed relatively low-key and local—in part because it's run by a co-op. But it isn't a logical lifestyle to hold on to. "Who among us would think of doing it now, and who would be able to pull it off?" Historian Annie Coleman posted that question when I asked her how she thought ski town life was changing. "It's been a long slow march toward increasing capitalization, and in a more corporatized labor market, the space for ski bums has been narrowing. It's racially and economically so narrow," she says.

The writer Wendell Berry once said that there are two kinds of people: boomers and stickers. The ones who keep wandering, and the ones who dig in. I am not proud that I've never made any place stick so far. And watching the ridge hippies hike up the boot pack together, or Shane trace the memory of past winters in the snow level, I wonder if I've done it wrong. There's something to be said for Marcus's dedication. Sticking comes with the knowledge of the ridgeline, where to drop and when to hold speed, a feeling that you belong.

PAMMY OR PETER PAN

At the bottom of Schlasman's, where she's stashed her backpack for the day, Pammy pulls out a pint bottle of rum and takes a swig. I gape until I realize she's just chugging water, because a reused Myer's bottle fits perfectly in her pack, but I wouldn't be surprised if she had booze tucked in next to her sandwiches, too. Pammy herself is pint-sized, a short person on huge skis, and her voice echoes when she laughs. She was an original ridge hippie, one of the people who pioneered the lines Shane skis these days, and now in her sixties, she's the ringleader of a pack of women who ski here every day, all day. They stash their lunch by the lift, and pee in the woods, so they never have to leave the mountain. They ski like thirsty teenage boys, jacked up on the terrain, rowdy and teasing each other. Marcus says Pammy is his unofficial mother, and when he introduces me, I am immediately infatuated. Pammy is the epitome of everything I idolize in a dirtbag. Tough, dedicated, immune to the stratification of society, and fiercely independent.

Schlasman's is a slow double with plenty of time for talking. You have to carry an avalanche beacon to get on the chair because the lift gives you access to a wedge of uncontrolled, slide-prone terrain. I snagged a ride with Pammy, and tried to ask her about how her life has been tied to Bridger, but she was cagey about her story, deflecting and laughing, clamming up when I asked about how things have changed over the years. That's another trait of ski bumming: keep the secrets close to the bone, protect your place and your people.

Pammy works as a baker for a guest ranch in the summer, which she's been doing for years, and then she's here all winter, committed to constant laps on Schlasman's. Her husband, Turk, is on the same program; they met here and they're still here. They're subversive and a little abrasive—no patience for someone like me asking them drippy questions about their lifestyle—and most of all eager to ski. Pammy and Turk are the deep-level dirtbags Bubba was talking about when we first met.

Pammy is sandpaper and she is magnets. Sun-cracked skin and a crackly voice. She barely waits for the rest of the group to slide off the chairlift and buckle our boots before she takes off toward the rim of Mundy's Bowl. We follow her gingerly on the rutted traverse, until we can drop down into the sun-warmed bumps below. It's one of my favorite types of skiing, big soft moguls with enough bite to edge into. The bumps funnel into gullies through the trees, and I follow the pack down into the basin, finding the sneak lines between cliff bands. Trying not to lose them in the rush. One of the women tells me it feels like dancing to her, this relationship with the mountains, moving downhill, finding the soft spots and the curves. She says she's searching to fill some kind of emptiness when everything else feels commercialized and too close. Being here lets her be wild and in her body.

The old ladies come every day, clicking into their skis like clockwork, even when it's crappy out. The rhythm is part of the beauty of ski bumming. Just show up for first chair, and keep rolling, assuming that you'll find someone to ski with. Do that all day, then slide into après as the sun sets. On the surface, leisure every day sounds like a dream, but someone like Pammy wouldn't be here if it was just about pleasure. There's psychological weight to the obsession, and a certain type of brain chemistry that connects here.

Kafka said we should follow our most intense obsessions mercilessly, and he might have fit in here. Skiing might seem loose and laid-back, but everyone I know who is a dedicated, lifelong skier has some kind of uptight obsessiveness around the sport, and an internal structure that dictates when and how they go to the mountain. In Utah, there's a guy called First Chair Brian who is number one in the lift line every day at Alta. Down the road from there, at Snowbird there's a pack of older dudes who have had breakfast together at the base of the mountain for the past forty years. There's a man named Mikey in Aspen who has skipped four days of lift skiing in the last ten years, and at Squaw Valley, KT Cheryl says she's only missed the first chair on a powder day a couple of time in the past thirty years, despite the hordes of young bros who try to beat her to the lineup. The obsession extends in so many directions. Benny makes his own skis, because no other ones are perfect. Todd, who I met here at Bridger, keeps a meticulously detailed ski and snowboard diary, noting conditions, crew, and anything out of the ordinary from that day. "Aprèspuncture" or "car in ditch," or "they said one inch but it skied like 10," some of his notes read. I recognize that need to compulsively catalog in myself. It's why I write.

One morning in Bozeman I have breakfast at the Western Café with psychotherapist Timothy Tate, who specializes in counseling young men who chase the Peter Pan fantasy of mountain athleticism. He cracks the yolks of his runny eggs over his hash browns, and tells me that he thinks the drive to push yourself in the mountains comes from a combination of genetics and soul. "The mountains are a perfect playground for that hero narrative in our youth-oriented culture," he says. "I began to see a pattern in men's lives. Their identity was based on their capacity to excel and conquer, and do what others haven't been able to do. The key word is identity, because their identity is so tied to their skill set, and the skill set is tied to youth and being a hard-ass."

Our self-perception is informed by what we do, Tate tells me. Our identities are forged through pattern: the people we see at breakfast, the places we go. If you can carve pattern into purpose, like Pammy and Marcus have, those individual tendencies can feel secure. But if your sense of self is based on a constant testing of boundaries, it's hard to find stability.

He says that it can be easier to dedicate your life to skiing than to a person, or a job, or the volatile calculus of figuring it all out. Tate says going to the mountains gives people purpose. You know exactly what your goals are, and the fear of pushing yourself makes you focus. It blurs the shape of the outside world. That singularity can be lonely, but you know where the edges of your aspiration are.

It's more than just that need to push and prove yourself, he says, and skiers often feel a spiritual connection to the mountains. The compulsion to be outside is grounded in a sense of awe. "We talk about a third variable based in philosophy and Plato's mythology," Tate says. "It's a force called character, and it's independent of genetics and environment, because those two don't account for the people we're talking about, who have some call that eclipses all other concerns, who say, 'I don't feel like myself unless I'm in the mountains.'" Plato called it the daimon, the unique carrier of your destiny, and depending on your preferred mythology or lack thereof, it could be called a spirit, or a genius, or a guardian angel. Tate calls it soul.

Tate has wild cotton-bush hair, and a bullshit-free smile, and when he talks about the intangibles, he hits on something I often struggle to explain. Sometimes I drive eight hours to ski for four. When the hot-breath panic of anxiety wells up in my chest, and I feel like I'm wasting my life, my first instinct is to run to the mountains and sweat through the sensation. I have no other broad cultural ties—no religion, no groups. I am skeptical of joiners, but I have skiing, which carries me around.

Jennifer Sofie Gulick, a licensed professional counselor based in Jackson, says she sees immature behavior—where people choose to focus on their own solo adventure quests, and avoid greater responsibility—all the time in the Tetons. She says it's an easy place to avoid growing up until, all of a sudden, it isn't. When people come to her, they're often struggling with depression because they haven't moved on from the life they wanted in their early twenties. They're creeping into middle age, their peers have kids and careers, and they feel stuck in their selfish ways. She says women often come to see her sooner than men.

Tate says the responsibility-shirking lost boy who never grows up, or never had to, is the most common stereotype of the ski bum. He (it's so often a he, although I don't exclude myself) is dependent on his parents, or his partner, if he can manage to hold one down. Selfish, risky, living in the moment, unconcerned about consequences or the long term. The presence and pervasiveness of those Peter Pans elides one of the core questions about the value of ski bumming. Is extending adolescence a recipe for a stunted, unsatisfying future, or is the singular focus of the lifestyle a kind of grace?

That's where it gets tricky for me. Nights when I tuck myself into my sleeping bag on a buddy's floor, after an evening of drinking locker room beers in my ski boots, I start to think too much about my own path. I worry that the ease with which I can traipse back into mountain towns, living off trail mix and Bud, is a sign of my emotional immaturity. Am I still fixated on this because I'm selfish and self-centered, and unable to grow up? Do I idealize people like Benny because his life seems like a positive way to spin my negative tendencies? I too easily identify with Peter Pan.

Skiing is something you do alone, even in crowds, and that solitude is both a defense mechanism, and a way to work through the world. Ski communities embrace a certain kind of loner. I know they've let me burrow into my own antisocial tendencies. I can flit through town, being with people without really having to connect; the common activity can be enough. And sometimes I think that's okay. Skiing gives you an unspoken way to relate, and a fragile but tough spiderweb of people who care about you.

SEEKING SENSATION

I left the ridge rats at Bridger one day and headed down Gallatin Canyon to Big Sky, the other ski resort in the area. Big Sky backs up to the private, high-dollar Yellowstone Club, where people like Bill Gates and Justin Timberlake have homes. It pulls in a different crowd than the nonprofit denizens of Bridger—fewer locals, more destination skiers—and the resort has been going through big changes lately. Big Sky took over neighboring Moonlight Basin in 2013 and became a part of the Ikon Pass collective in 2018, which brought in a wave of nonlocal skiers. But in spite of that creep toward corporateness, or maybe because of it, Big Sky has long been known for its gritty, raucous local ski bum community, skiers shaped by the edgy, shark-finned mountains of the Madison Range. The locals crown a dirtbag king and queen every year and have since 1979, and the parade of past kings and queens is very present. That royal court is comprised of postal workers and snowcat drivers and ex-racers who are better than you at both skiing and slamming beers. The Big Sky version of chasing the dream is more amped up than Bridger's.

That first morning in Big Sky I go straight to the top of Lone Peak with Joe Turner, a local horse trainer who skis here almost every day—business is busier in the summer, he says—and who has a horseman's steady cool and quick twitch reflexes. We are blowing cold air through our noses in line for the tram. At the top, spiky and shrill in the cold, we stare down the sheerness of the Big Couloir and the Little. You have to check in with ski patrol at the top, because they only let a certain number of people down each run at a time. We wait in the patrol box for our line to clear, then slide out through the ice-sheared gate, where they'd told us to be careful. We take a couple ginger turns up top, linking loud ice patches. Then the flanks of the chute open up, creamy and smooth, loaded with enough snow to push back on your edges, and spring you into your next turn. There is purple, early morning, midwinter light, and just-softened snow. A few careful turns down the gully, then the open-up-and-let-'em-run pitch of the bowl. That perfect physical balance, a spike of adrenaline that gives way to the grace of the float. The catch and release I'm always chasing.

If your brain chemistry is set a certain way, you truly are always chasing that sensation. The more you ski, the more the ache for dopamine (the neurotransmitter associated with motivation and pleasure) in your body keeps rising. You can become addicted to the feeling of risk and reward and that can seep into so many aspects of ski town life. Maybe you party too hard, because you are good at extremes, and because everyone around you is on vacation. Maybe you see people turn to harder drugs when they blow through their limits, or when they get hurt and have to cope. Or maybe you're compelled to push harder and farther into the mountains, even in the face of trauma and loss, because your body is the one thing you can control. That churn of adrenaline is exhausting, but you—we, because I do it, too—still call it living the dream, even when it's destructive. That's the dichotomy Tate is always trying to untangle, and to teach his patients to live with.

So when does obsession tick over into obsessiveness? And how much of it is governed by biology? What is science and what is soul? Psychologists have found that the most significant common trait among people who are pulled to the mountains is something called sensation seeking. "Neuroscience calls it novelty, it's the willingness to take risks for the sake of rewards," says Cynthia Thomson, a health researcher who looks at behavioral genetics. "Compared to people who don't ski, skiers are higher on the sensation-seeking scale. They have a low threshold for boredom, and tend to look for more exciting experiences, even if there are dangers associated with them." Thomson was interested in sensation seeking because she's a skier, too. After spending a string of seasons at Whistler Blackcomb she wanted to understand why some of her friends couldn't break away, and why she kept pushing herself in the backcountry. She's learned that sensation seeking can express itself differently depending on the person and the venue—she says party culture is tied into sensation seeking, as are gambling and unsafe sex—and in the mountains it often comes through high-consequence ski runs.

In 1964, University of Delaware psychology professor Marvin Zuckerman created the sensation-seeking scale as a way to quantify the trait and its correlation to risk. Sensation seeking is a combination of four traits, which show up variably, but are related. The first is what he labeled thrill and adventure seeking—it's the physical pursuit of excitement. The second is experience seeking, which is the mental stimulation that comes from being in new places. The third is disinhibition, the inability to follow socially appropriate behavior, and the fourth is being highly susceptible to boredom.

It's never just one factor, or one part of your brain that lights up when you're taking risks, and skiing can hit on all four of those traits. Everyone's personal brain chemistry, body awareness, and history is different, and genetics isn't a clear indicator—my own brother, for instance, is an economics professor, whose job consists of carefully analyzing risk from the confines of a classroom—but there are some common biological traits among ski bums. Sensation seeking, for example, which is also connected to traits like neuroticism and being tolerant of chaos, tends to show up early. Small kids exhibit it on the playground, they're the ones launching themselves down the slide, bloody kneed. That appetite for risk can "decrease with age, but if you rank everyone, that comparative ranking tends to be relatively stable over time," says Dr. Scott Carlson, a professor at the University of Minnesota who studies risk. If you're a sensation seeker, you'll always be one.

Carlson says it's hard to determine what makes some people risk takers, but there are personality trends that tie together sensation seeking, impulsivity, and the need to live on the edge. His research is focused on identifying the underlying biological and environmental factors, and the commonalities that cause certain types of behaviors. He wants to know whether it's nature or nurture or neither or both. One of the traits he sees the most are those antisocial tendencies, which are connected to both sensation seeking, and rule breaking. At its far end, antisocial behavior is correlated with sociopathic tendencies, and a full disregard of social norms. It's connected to binge drinking and binge eating, borderline personality disorder and psychopathic behaviors, but it also exists on a spectrum. And he thinks a lot of skiers might live in that place on the spectrum where they need to push, and where the rules don't feel like they apply. "Most people hear that word *antisocial*, and think it means people who don't socialize well, but really it's people who have a pattern of chronically acting against the norms of society, and a higher frequency of rule breaking," he says. "They're iconoclasts, which might be good in some ways."

Once Carlson mentioned that idea of antisocial tendencies, I considered most of the people I've met in the mountains. I found it hard to find one who didn't show them to some degree. It's Benny flipping his jacket inside out to break the rules, and

Pammy's obsessive everyday mission. It's Leger calling bullshit on his high school career counselor. I'm not a psychologist, and I don't want to throw broad assumptions over anyone's brain, but it's normal in the mountains to not want to be normal.

If you're someone who seeks sensation, and doesn't abide by social norms, it can be hard to know where the line between ability and risk might be until you're way out past it.

One day on the Bridger ridge, I drop over a different side of a rock from Marcus, Travis, and Todd and suddenly they are gone, and I'm alone in a spiny web of chutes, scared of what's below me. I'd been posturing, pretending to be able to keep up, and once I lose the training wheels of their presence, I am unsure of which chutes close out and which I can ski down. I know that when you look at a map of the ridge it's more rock than snow, but I'm not sure where those cliff bands might be, because I'd been following blindly. I get shaky. I yell for them, and second-guess every turn. I end up pinned on a band of rocks, unsure of how to slide through the skinny channels of snow. I try to backtrack up the gully, sidestepping and grasping at tree branches, until I'm sweaty and near tears. Eventually, I find a traverse that trails into an open gully, and skitter down, scared and embarrassed. When I get through the chute I'm angry and shook up, exhausted by my own fear. Part of why I'm no longer here all the time is that my need for dopamine has gone down. I don't love the clammy hand of adrenaline as much as I used to. I get scared more now.

It starts in the lungs for me, the vascular contraction of bad ideas. I get short of breath, and it's usually worse if I've been hiking, chasing people who are faster, fitter, smoother, and more confident than I am. From there it moves out like a star. I get shaky in the knees, like my blood isn't circulating enough, even though my heart is pounding harder than it should. It seems like a biological flaw. I don't like fear, exactly, but I find myself seeking it often, because I don't know how to stop myself, and because I want the elemental neuron massage that comes from the float and flow.

Aside from sensation seeking, risk has always been a part of skiing. Doing something hard, scary, and potentially destructive is a way to feel very alive, and you can literally become addicted to the feeling. Your body begins to need more dopamine, and certain people are more inclined toward the excess than others.

Jackson Hole–based addiction counselor Ryan Burke, who founded the Mindstrength Project, which helps mountain athletes deal with stress, told me that dopamine is a wanting chemical, and its curve in your brain is an asymptote. "You come really close to getting that satiated contented feeling, but your body makes you want to go back for more," he says. The more you get the more you crave.

Scientists used to think that addiction was a character flaw, but thanks to modern neurology, there's now a leading theory called the brain disease model of addiction, which says that addiction comes from both genetics and environmental factors that can change your brain.

Dopamine is our reward system. It's the thing that makes survival, eating, and sex feel good. When you hit a pocket of perfect snow, and carve a weightless turn, it lights up those same neurotransmitters, sending a chain reaction to your brain, telling it to try to find that feeling again and again and again. Anything that's addictive is so because it triggers a dopamine response. It keeps us motivated, but when you mess with your dopamine system you can get hooked. "You don't want to flood your body with any neurotransmitter," Burke says. "But it's so easy to jack our dopamine receptors these days. We have synthetic drugs, we have skydiving, we have casinos, and we have skiing."

Burke says that when you elevate your levels of dopamine outside the norms that our brains are adapted for—by doing something like skiing—your dopamine levels rise, and then they drop down lower than they were before, because your brain predicts you're going to get another big dose. Your baseline changes and it's hard to get back to homeostasis.

Burke says it becomes an issue when you start prioritizing skiing and its accompanying rush over everything else. "It gets bad when you're like, 'If I don't have this thing then life isn't good,'" he says.

Certain people want that hit more than others because everyone's mesolimbic approach system (the pathways that pump dopamine into our brains) are an overlay of several different factors. According to David Zald, a neuroscientist who studies dopamine, your reaction is based on the genetic configuration of auto receptors in your brain, and on your number of reuptake sites, which drugs like cocaine and amphetamines, as well as many common antidepressants, attach to. But Zald says it's not strictly based on how you're wired, it's also determined by your past experiences. "When you get to the extreme ends of psychopathology I can't put someone in an MRI and say, 'Ah, based on this brain pattern this person is going to be an extreme risk taker,' but there are patterns," he says. He's seen that people who are more likely to be impulsive, and need greater thrill and novelty, have dopaminergic systems that are less controlled. "If dopamine is acting as the gas, pushing one toward risk, they can't put the brakes on as well," he says. It's why you might get back on the lift after freaking yourself out in a fall. Your brain can't cut off the source, even if it should.

We all have a dopamine D2 receptor gene, which regulates how much dopamine is naturally fed to our brains, but they're not all the same. According to studies from the University of Florida, about two-thirds of the US population has a "normal" form of D2 receptor circuitry. But that means that there's about a third of us who are in limbic limbo, who might need more dopamine to feel excited, which is known as reward deficiency syndrome. Those are the people who are more likely to forgo work for powder mornings.

Having a certain brain chemistry isn't going to specifically predict your life path, but those receptors create patterns. "If you're someone who is a ski bum, you might have fewer receptors, so you need a higher high," Burke says. "Some people can go to Disneyland and have fun, you need to go to Disney on meth." Couple that reward deficiency syndrome with the unnatural cultural reinforcement of being in a ski town—where you're socially rewarded for high highs and risky dopamine-inducing behavior—and it makes the spikes even more dramatic and necessary inside your brain. Even when I am scared following Marcus, Travis, and Todd, when I get to the bottom of the run, the rough parts have faded, and I slide into the lift line again.

We know from a range of research that sports are good for your mental health. Moving your body benefits mood, concentration, sleep, and self-worth. Exercise is connected to lack of depression, and it can also be a coping mechanism for unchecked anxiety. Tate says he can see that trend in younger guys when they move into the mountains. They keep going higher and farther, skinning up mountains before and after work, running for miles in the summer, shoving whatever doesn't feel good aside. "They have exercise, which is probably the second-best drug for mental illness, because it makes you feel better," he says. Skiing all day might mean you're training for something or it might mean you're fire-hosing your anxiety. The internal feedback loop of D2 receptors encourages the behavior to continue, even if it's unsafe. "If you get on top of a mountain, then you have a story to tell, but there's a relentlessness to that," Tate says. "Once you achieve something, what's next?"

Now, whenever I visit a ski town, I see the sliding doors of my own life. I can consider the paths of the people I worked with when I was in my early twenties. One became the mayor of Dillon, Colorado. Two of them started a business selling ox-ygen to tourists who aren't acclimated to the altitude. It's doing well. A lot of the guys who stayed are still patrolling. They've taken on more responsibility and their aims are different. Today they're carving out a life in tough-to-sustain tourist towns, try-ing to have kids, treading water. Now, in our midthirties, we calcify, stuck in our ways, our bodies not as flexible and resil-ient as they used to be. Habits are harder to break, so are the human connections, the sense of community. From the outside I can still see the appeal.

To the dedicated few, skiing is as vital to life as art or song. It fills in the gaps of your existence with fire and calm, soli-tude and connection. But, like Leger told me in the tram line at Jackson, skiing may momentarily meet all your needs, but it meets almost none of them in the long term. Very few people can live powder day to powder day without more of a purpose for long. To make it feel good forever, as someone like Pammy has, it must continually feel good, whether due to your brain chemistry or the strength of your community.

I believed the myth of the ski bum, in all its weirdness and brokenness and selfishness, because I wanted the dopamine and the belonging and the feel of it: the romance of the early morn-ing sparkle-studded glint of the hoar frost, or the way I forget to breathe on bump turns. "It sounds hokey, but when I'm skiing everything feels like it's in balance and synced up," Marcus tells me on a ridge hike, when I ask him why he's pegged his life to skiing. "Everything else melts off, everything makes sense, and everything is smooth."

Skiing is the closest thing I know to flying. I don't know if it's dopamine, or desire, but when we stop on the ridge at Bridger, as I steel myself against the wind, I can feel the good kind of adrenaline coursing through. Tips pointed into the void, my body knows what to do and I drop.

SECTION 4

OUT IN THE COLD

Who gets to be a skier

SKIER TRASH

Warren Miller invented the ski bum dream, but he never said who it was for. The glitz of the movies, and the gloss of resort towns, ignores the grit of reality. Miller created a narrow archetype that persists today.

If you look closely you can see the cracks in the ski bum ideal. Part of the myth of dirtbagging is that you ended up in the mountains by some kind of cosmic force. That the great gods of powder put you on the chairlift with Benny and Leger. That you were somehow chosen. A cultural feedback loop reinforces that story—and it does so for a select few.

Skiing is exclusive on so many levels. Even if you have the personality type to push your limits, and the internal hunger for dopamine, you need the opportunity to be able to do so. It's a narrative problem and an ethical one. If you look around on any given day, at any given ski hill, the snow is very white and so are the people. More than half of regular skiers earn annual incomes of over $100,000, and eighty-seven percent of them are white. When you look at ski media, from the yearly stoke flicks to the Hollywood films, the protagonists are almost exclusively straight, able-bodied men. Since the days of the 10th Mountain Division, nearly all the stars in all the ski stories have been white dudes with enough cushion to get risky. If you're looking for a hero who looks like you, and you don't look like that, your options are limited or nonexistent.

Skiing is broken because of that narrowness. The financial barriers to entry, like gear prices and private lessons, are steep especially if you try to stick around and make a life of it. The social barriers of embedded racism, ableism, and sexism are worse. What I offhandedly call ski culture is couched in so much: race, money, gender, what you wear, how you dress, and the ways you interact with people. The carefree soul skier living close to the earth is a carefully codified stereotype, and the carefree part often comes from not having to police your language or your appearance, and the privilege of implicitly feeling like you belong.

It's easy to fit in when almost everyone looks similar to you and has analogous past experiences and knows the same slang. The only thing I don't have on my side as an able-bodied, white, straight, financially-cushioned female is the female part, and even that sometimes feels like it makes me special instead of sidelined. The fact that I can get in my car, head to the mountains, and assume that I'll be welcome is just one in a long line of opportunities that have flowed my way, some warranted, some purely luck, starting with learning to ski before I can remember. I got a lift from the freebie high school ski club, but I also had parents who encouraged physical risk. Who believed that something like skiing was valuable and who had the means to put money behind that belief. Because of my upbringing, I have always assumed that if I show up somewhere, and can ski fast enough, I will eventually be accepted. But I know that things would be different if I weren't white or if I didn't come from a base level of privilege. If I presented differently, would people be so willing to ski with me, and spill parts of their life when I rolled into town asking nosy questions?

I have been asking a lot of those nosy questions about intent and desire and risk, but it's harder to ask the ones about identity and privilege. Do your parents help you out? Do you have any friends that aren't white? Would you date a girl who was a better skier than you? Why aren't you washing dishes or working housekeeping if you say you're pinched for money? This line of inquiry is uncomfortable but necessary if we want to understand the current state of the sport.

To even begin to ski you need enough expendable income to invest in gear, gas, and leisure time. Kitting out a kid costs hundreds of dollars, and a holiday weekend of family skiing can add up to thousands. Skiing is one of the most expensive pastimes in the world, somewhere between polo and golf. Lift ticket prices are constantly rising, and they're just part of the sunk cost of transportation, lodging, and food when you're stuck in a resort village.

The sport has evolved to a point of precarity because of those barriers. Once there were more than 600 local hills scattered across New England, many of them community-run rope tows with pocket money day rates. Now there are fewer than sixty, and that winnowing is true everywhere. I spent a chunk of one winter in the scrubby mountains of West Virginia, skiing the trees of Snowshoe Mountain and Timberline Ski Area with a group of fervent skiers from the Canaan Valley. The wrinkled terrain of the West Bygod is some of the most mountainous in the country, but it's tightly treed, and low altitude, which makes for tricky skiing. The snow was sloppy. We tracked through the remnants of wet storms turned to rain, surfing slush puddles by leaning back on our skis. I followed Bob Lilly, who grew up close by and started a brand and ski crew called Skier Trash (they "sponsor" freestyle skiing icon Glen Plake), which mainly produced T-shirts, videos, and hometown pride. Bob's ski style is one part park rat, hitting every possible jump and feature, one part '80s-style freestyle hotdogger. He throws tip rolls and worm turns and other antiquated tricks, because, he says, in the scrubby Appalachian mountains you have to make it fun for yourself. "Got to take advantage of the flats around here," he says, swirling into some ski ballet. Bob thinks he loves it more because he had to scrape and scrap, fishing for hand-me-down gear, wearing whatever outerwear he could find. Skiing the streets when lift tickets weren't attainable. He chose skiing on purpose, instead of having it handed down to him. He lived in Lake Tahoe for a while, but he moved back to his home mountains because he says it felt important to protect and encourage this rugged little ski scene. Bob pointedly skis in tight black jeans, even in the rain, and he has the words *skier trash* tattooed over his heart. He wants skiing to feel punk rock and accessible, and he wants to shine a light on an increasingly rare place like this where you don't need to be rich to ride.

There are still community ski swaps where you can exchange used gear. There are free skiing programs, like the one for all Vermont fifth graders, and there are nonprofits that try to encourage diversity in the sport, or to give low-income local mountain kids a boost. Some of them have a wide spectrum of use and extravagantly wealthy funders, like the Doug Coombs Foundation in Jackson Hole, which works hard to get the children of immigrants working in ski towns out on the mountain. But nonprofits aren't a sustainable way to democratize skiing. Even if they help kids fall in love with skiing, they can't dismantle the sport's inequality. You need access and someone to bring you to the mountains, but then you need constant money and free time for the tickets and the gear. Bob has tried to make a mission out of being an outsider but it's hard, if not impossible, to break into a judgy, image-based sport if you can't afford to. Bob is rightfully pissed about the boundaries of the ski world because he feels like he falls outside of them. It's hard for a kid growing up in Beckley, West Virginia, one of the cheapest places to live in the US, to find a place in the ski world, even if he loves it with all his heart.

AS WHITE AS SNOW

Financial privilege can be sneaky and obtuse, while racial privilege is often more blatant. The skier population was overwhelmingly white at the outset, and not much has changed since the early 1900s. As of the early 21st century Black skiers made up only two percent of the US skiing market.

Sociologist Anthony Kwame Harrison grew up skiing in New England, and the sport is a big piece of his family's story. His grandparents were some of the original investors in Black Mountain in Jackson, New Hampshire, where his grandfather had the ninth-ever season pass. He says the sport's whiteness is ever present, he's noticed it since he first stepped into bindings as a tiny kid, but skiing's undercurrent of racism really clicked in when he moved to Virginia for a job, and started frequently skiing the mountains of the Mid-Atlantic. He says that on nearly every chairlift ride he would get questions because he is Black: How are you such a good skier? Who are you here with? What's your deal? The curiosity was constant, and sometimes pointed or hostile, and it underscored the sport's lack of diversity.

That questioning on the chairlift threads together Harrison's personal life and his academic interests. His research spans psychology, philosophy, marketing, and anthropology. In his work unpacking racial identification and social spaces, he's dug into the whiteness of skiing, and the ways that leisure spaces hold racist boundary lines. He quantifies the harmful but pervasive stereotypes in the sport—that Black people don't like to ski, or that adventure sports are exclusively for white people—to show the past and current context for skiing's glaring whiteness. "It comes from a time when white men ruled, and the world was theirs, and the frontier was theirs," he says. "As other voices are rising and claiming identity, there's a sense of anxiety, and sometimes that manifests in real dogged and ugly efforts. Some of it is people who don't think of themselves as racist, and don't want to be bothered with those things on vacation, but some of it is a sense of a shrinking core American whiteness and male authority. It's the same as the rise of the alt-right movement and the things that Donald Trump got away with saying."

Long before white nationalists took to Reddit and Trump was voted into office, ski towns were constructed, whitewashed communities that mimicked ski villages in the Alps. Early resort owners who wanted to glom on to the classy European image brought in Austrian ski instructors and gave ski runs names like Vista Bahn and Trans Montane. Everything from the food to the architecture was Eurocentric and implicitly white. Skiers still use the phrase après, even when we're just drinking beer in a parking lot.

The vernacular was a mashup of European heritage and the very American notion of redlining based on race. Harrison says the bone structure of a ski town was and still is a utopian idea created by white people who want to feel like it's "their" place. They don't want to be uncomfortable on vacation, and often they have the ability to box out any awkwardness, he says. Being able to avoid thinking about something is the embodiment of privilege and for white people that is deeply embedded in American land use, social structure, and politics.

Harrison's paper about the cause and effect of privilege, "Black

Skiing, Everyday Racism, and the Racial Spatiality of Whiteness," points to the fact that in the US, aside from the South, most of the settlers who moved to mountainous areas were white. It digs into a deeper history of environmental racism in the geography of exclusion. It's necessary to consider how European settlers took over the land in the first place, ravaging health, culture, and ecosystem as they claimed the continent for their own, confining indigenous people to reservations, and co-opting their landscape, language, and livelihood for their own gain. Even the names of ski resorts—Arapahoe Basin, or the cringeworthy Squaw Valley—blatantly point to what was taken.

The conception of recreation in the US has pretty much always revolved around what was convenient or relevant for white people. In the late nineteenth and early twentieth century, when the predominant cultural idea of wilderness shifted from an unknown scary threat to a sublime fantasyland of adventure, it was often placed in contrast to crowded cities filled with people of color. For white people, the mountains were an escape; for minorities, they were often filled with threats. "Would you want to go into the backcountry if it was historically associated with lynching?" Harrison asks me, underscoring the difficulties around changing the cultural coding.

As resorts grew from the initial small alpine village model they became escapist havens for second home owners. As far back as the '30s skiing was about getting out of the drudgery of the city. Harrison says that in the post–civil rights movement, resort towns became places where white people could get away from the unpleasantness of politics, and that exclusivity has stuck. To show the deep roots of racism, and how they present now, Harrison unpacks the lazy arguments around lack of minority participation. Those include the "marginality hypothesis," which argues the high cost of skiing prices minorities out. That assumes a social class correlation tied to race, but really there's no statistically significant spending differences between Black and white vacationers. Skiing is expensive for everyone. He also points out the flawed theory that people of different races have fundamentally different interests. Instead, he says it's a pathway problem.

There are subtle markers that delineate space, as well as clothing, language, social codes, and behaviors that indicate someone might be an insider. I often make immediate judgy assumptions based on what other skiers are wearing, or how they carry their gear. If you look at the symbolism and the cultural clues about who is welcome—the athletes, the ads—they're largely white. When you show up to ski, you're facing everyday systemic racism, as well as the factors specific to these towns. Sociologist Pierre Bourdieu calls those signals "hidden entry requirements." They're the secret handshakes that show you belong, which also build invisible barriers.

Harrison says that if you live outside of those boundaries you often need someone to bring you in and make you feel welcome. You need intention to go skiing. It's hard, expensive, and remote, so someone who doesn't have a connection, or who has a bad experience, might not want to come back. "It's hard to make that investment of time and money," he says. He's had his own rough experiences, where he's been targeted and judged. Sometimes it's subtle enough that he can't tell if he's been singled out, but it still puts him on edge and makes him feel unwanted, like when his skis were stolen on his first trip to Killington. "Maybe I was targeted, maybe I should have just locked my skis," he says. "But if you're already not sure if you belong in a space, a very small gesture can push you away."

There's a strong community of Black ski clubs in the US, capped by the National Brotherhood of Skiers, which has been hosting ski summits since 1973. That year, after two years of planning, Art Clay, the trip director for the Sno-Gophers Ski Club of Chicago, and Ben Finley, president of the Four Seasons West Ski Club of Los Angeles, hosted the first-ever Black ski summit at Aspen, to give groups like theirs a place to gather and connect, so they could all feel comfortable on snow. 250 people from thirteen different ski clubs showed up, and since then they've continued to host annual summits. Resorts now vie to host these summits, which are regularly among the largest ski industry gatherings, and which bring community and a heck of a lot of commerce into ski towns.

"It's like a snowball. It started rolling down the hill in 1973, and it gets faster and bigger the longer it runs," Finley says. "Our challenge was dealing with history, and causing our own people to realize they could change their lives by skiing and boarding."

It's one of the largest ski organizations in the country, but Mike Lanier, who ran the NBS's youth program, says that doesn't mean it fit fluidly into the rest of the ski world. He says it was consistently exhausting to try to funnel kids into ski programs and race academies, especially in places where there was no diversity at a high level. Harrison says the clubs tend to encourage periodic Black skiing events rather than a regular and sustained presence on the mountain, and even though it's a way to introduce people to skiing, and create camaraderie, it can also contribute to the othering. Finley and Clay were inducted into the Ski Hall of fame in 2021, but Finley says it only happened after three tries, and they only got the nod after Olympic Alpine skier Billy Kidd advocated for them. "The best part of us being inducted into the hall of fame is that the NBS is now a voting member of the ski hall," Finley says. "We have a seat at the table that is extremely important."

The lack of a seat is a systemic problem that's hard to shake, especially in opt-in places. "Certainly there are bigger concerns about recreation, but it's the idea of creating exclusive spaces to not have to confront a diverse society. It contributes to people being invisible," Harrison says.

To try to counter that exclusivity and render non-white people visible, Stan Evans, a snowboarder who grew up in Alaska, has spent his whole career pushing, often painfully and loudly, for inclusion. Evans has been shooting ski and snowboarding since the late '80s and he says it's been heavy to carry the weight of both being the only Black person in the industry by virtue of his presence, and of trying to increase representation of non-white athletes through his work. Representation is crucial, both in the ads, marketing, and athlete images that we see, and in the workplace. He's creating role models by both paths. He feels the pressure to go above and beyond, and lead the charge so that other folks can come up behind him. "They need to see people like me succeeding," he says. "I worked my ass off. I worked ten times harder than anyone else. I had fuel for the fire any time anyone made an assumption that I didn't know what I was doing because I was Black."

To make skiing equitable we have to address racism on every level: social, spatial, economic, environmental, and political. Harrison says that in his experience, being the only Black skier on the mountain can, in a weird way, make white people feel more comfortable. "It's like, 'Hey, this guy is here, so we must be OK,'" he says. Evans echoes that sentiment, and he says that's a big piece of why he's still one of the only Black snow sports photographers. No one is seeking out other shooters or filmers. He's trying to pull in the next generation, and to visually show non-white athletes, but it's exhausting to do that alone.

Harrison thinks the tokenization makes people feel like they have permission to ignore the deeper, dangerous disparities, especially in their leisure spaces. "It really addresses communities who like to think of themselves as woke or progressive, but shows how blind you have to be based on the kind of world you create around you."

He thinks that diversity is only acknowledged or accepted in a very narrow way, which is small-minded, dangerous, and stuck in the past. It's never going to change until we consciously create a better world, and acknowledge why this one doesn't work and isn't fair. Harrison says the skiing crowd, which is mainly made up of people with some level of intellectual privilege, should know better. "There's no fair excuse for ignorance," he says. As in any space, the onus is on the majority, white people, to acknowledge embedded racism, and make changes. We need better pathways, to change the shape of the story. Lanier says he thinks it's absurd and reductive that the ski industry is so homogenous, and it's killing the sport. Skiing will never be good, and it will never grow or sustain itself, if it's only white. "If you're really concerned about the sport, you've got to introduce people," he says. "You're not going to get everyone, but you have to open the doors."

FRONT OF THE HOUSE, BACK OF THE HOUSE

I spent several seasons of my ski town life serving pizza, and at the restaurant, the people in the front of the house—the waitresses and hostesses, the bartender and the pizza cooks manning the big showy oven—were the usually-white skiers and snowboarders who showed up for the dinner shift after their days on the mountain. In the back, behind a muraled wall, cut with a skinny window for popping through plates, Caesar ran the kitchen. The guys in the back doing prep work and most of the actual cooking were mainly Mexican, and not, as far as I knew, skiers. I'd chat with them in my measly Spanish, but we didn't talk much about our lives, or how we ended up there, passing plates across the wall.

That wall is a too-obvious marker of one of the deepest unspoken divides in skiing. It's the gap between the people who present as skiers, and the ones who are doing the grunt work that keeps tourist towns turning: flipping sheets in hotels, or frying chicken nuggets and tots for lunchtime in the lodge. Ski town economies rely on low-wage workers to survive. In a lot of places if you look past the novelty service workers—the white kids like me who could make skimpy paychecks because they were really there for the skiing—you see the people who are in those towns because there are plentiful jobs, even if they don't necessarily pay well.

The population on the trails and chairlifts is heavily white, but if you dig a little farther into most mountain communities, you'll see that they're much more diverse. In big ski destinations like Jackson Hole, Mammoth, or Telluride—or Eagle County, where I was slinging pizza—the towns are actually about a third Latino, give or take a margin of census accuracy based on seasonality, and the fact that some of those workers might rather not be identified by the government, for the sake of their immigration status.

It's getting more expensive to live in ski towns, and wages are staying flat, which makes it harder to fill service jobs. "A big part of the reason is—most Americans want a year-round job. They don't want a revolving-door-type job," Dave Byrd, director of risk and regulatory affairs at the National Ski Areas Association, told the *Tahoe Daily Tribune*. A 2020 National Ski Areas Association report found that half of US ski resorts were having trouble filling jobs. But some of that trouble is due to the pay scale, and who can and will work for less than a living wage. According to census data, per-capita income across the board in Jackson Hole is $39,300, while the average for those who identify as Hispanic is only $14,400, and they make up nearly a third of the population, both in Jackson Hole, and on the other side of Teton Pass, in Driggs, Idaho, where the cost of living is a bit lower.

Many residents of the Tetons come from San Simeon, Tlaxcala, a small town 100 miles east of Mexico City, with a sluggish economy. In 2019 Jackson-based immigration lawyer Elisabeth Trefonas (one of only three immigration law experts in the state of Wyoming) told the *Jackson Hole News and Guide* that up to eighty percent of the people she sees come from that area. In the mid-'90s a group of workers started coming up from San Simeon, because they could find well-paying jobs in Jackson where the restaurants, hotels, and construction sites needed workers. They can earn up to ten times what they can make in San Simeon, but a lot of them want to go back to Mexico.

Hispanic immigrants who are already struggling to stay afloat on low-wage services jobs in expensive mountain towns also face the cut of other micro- and macroaggressions. Reports from local social services organizations in Jackson say that people also feel like they're illegally kept out of housing, and discriminated against for jobs. And because of their immigration status it's hard for undocumented workers to get access to the kind of human services and resources someone might need when they're working for a third of average wage in an already expensive town. It's always been hard to sustain that life, but it became even more precarious during the Trump presidency. In 2019 ICE came through Jackson nearly every month. Trefonas told the paper that she'd been seeing more families and individuals without criminal backgrounds being detained than ever before. Kids stayed out of school, people didn't leave the house. And in the wake of those scares, restaurants and other businesses had to close, because their employees were either rounded up, or scared to show their faces.

The threat of deportation, and how it impacts businesses, illuminates how many people in those towns are undocumented, and how crucial they are for the economy. And it's not just Jackson, but many mountain towns. In Eagle County, thirty percent of the population is of Hispanic descent, and according to the county, regardless of immigration status, they're more likely to work those low-wage jobs, and have less job, housing, and income security.

The truth is that a solid and growing portion of ski town residents are people like Javier Pineda, who came to Summit County, Colorado, from Pátzcuaro, Michoacán, Mexico, in 2006, when he was 12. He joined his father there, and enrolled in a school system that is now around 40 percent Latino. Javier was Student Body President of Summit High School, and a Boy Scout. He now works full time as a paralegal while he's wrapping up college. He's also undocumented. In 2013, he became a Deferred Action for Childhood Arrivals (DACA) recipient, along with 15,000 other young people in Colorado alone. That means he could safely remain in the U.S. for at least two years. It also meant he could attend college in Denver, be eligible for scholarships, and get a driver's license. But in 2016, when Trump was elected president, and rescinded the Dream Act, that safety evaporated for Javier and his fellow DREAMers. Javier moved back to Summit County, and helped start a non-profit called Mountain Dreamers, which provides legal counsel community services and immigration information for undocumented folks in the community. He's decided to be public about his immigration status, because he knows other people can't be, and because he's seen how deeply immigration is entwined into ski town living. It's folks like his family, who have now worked Breckenridge for years, but it's also the big resort, construction, and food service business that provides so many of the jobs in the community. The face of towns like Breckenridge is changing, and Javier is trying to make sure it's happening in an equitable, true way. In 2019, he rode his bike from Summit County to Aspen, stopping at schools and community centers in mountain towns to talk about his path, and to advocate for the DREAM act and other pieces of immigration policy along the way. He knew he was putting a target on his back, by bucking stereotypes (brown kids don't road bike) and by publicly talking about his and his family's immigration status, but it felt more dangerous not to.

The story he's telling—that towns like Breckenridge depend on and exploit undocumented immigrants—is not in line with the nostalgic story of the ski bum, but it's an essential, true part of the narrative. And one we have to address and do better with, going forward. I'm not an immigration lawyer, or any kind of policy expert, but I can see that those towns can't sustain themselves without an immigrant community—the impacts of even minimal deportation would be devastating on the local economy. Young adults like Javier are making places like Breckenridge healthier, more equitable, and more robust. Lack of respect or recognition for people who are actually doing necessary work is an insidious problem in the broader American economy. By propagating the story that ski bums are the ones doing the shit work in ski towns, we're painting a fake picture of what those places really look like and who lives there. And by there, I mean America.

THE ODDS ARE GOOD
BUT THE GOODS ARE ODD

My friend Sheldon Kerr says she's getting her PhD in ski bumming, and I think she's right. She's on her way to becoming one of only twelve female certified International Federation of Mountain Guides Associations guides in the US—she's literally one of the best people in the world at what she does, but to get to the top she's had to battle a range of unseen and insidious obstacles.

We met back when we both worked for an organization called Babes in the Backcountry (I'm sure the name didn't help) which ran avalanche classes, clinics, and backcountry ski trips for women. I was just out of college, and Sheldon wasn't even done yet, but while I was the lackey who schlepped gear and made sure everyone signed their release form, she was teaching classes and leading women three times her age.

I caught up with her recently when she was in Jackson Hole, working for Exum Guides, one of the oldest, most respected guide services in the country. When I told her I was coming to town she said she'd take me out to ski something scary. "We're going to be the best ski mountaineering feminists in the universe," she said in her typical tone of big-hearted sarcasm, as we walked out to the top of a couloir called Chuter Buck. She hummed and cracked jokes as she built an anchor and calmly told me how to rappel down into the chute on my skis. I was saucer-eyed and shaky-legged, and she was literally singing, barely even breaking a sweat as I begged for snack breaks and rest. She's incredibly good at what she does—amazingly so in my eyes—but in a dude-dominated mountain guide world she's still fighting against stereotypes. In becoming a full guide she has the added burden of both trying to carve out space, and pulling other women up beside her.

Gender is another fault line in skiing. It signals who is welcome and who stays. Who can lock down jobs, who might get paid to ski, who finds mentors and who gets marginalized. Who is comfortable skiing alone and drinking in a crowd. It's hard to bro down and break in, when you are not, in fact, a bro.

Sheldon says that yeah, sure, she is physically different than the jacked dudes who are on the same career course as she is. But the hardest part isn't the physical part, it's finding the opportunities for learning and promotion when she's judged differently at every step. She has few female peers, and even fewer female mentors, and as she progresses there's the added pressure of having to be a role model. She wants to provide that broad mentorship she didn't get, but it's exhausting. The lack of female mentorship manifests itself in subtle, deeply ingrained ways, ones that easily push people out. "It's such a physically intimate sport, and men don't ask women to go on expeditions because there's a level of it being inappropriate," Sheldon says. "Think of how many dudes had older dude mentors and how strange that might be for a girl. I don't see myself having the same kind of informal learning opportunities."

Those informal opportunities, like the way a younger guy might be taken into new terrain by an older dude, are how people ascend in the guide world, and really any world. A lack of representation makes the sport less accessible for women. Sheldon's trying to reverse the trend by arranging all-female expeditions. She also started an inclusivity committee within the American Mountain Guide Association to formalize mentorship, and address the subtle and insidious sexism within the guide certification process. "The microaggressions are so micro," she says, noting how assertiveness is heralded in male guides but tracks as bossy in women. Sheldon knows that older guides probably aren't intentionally excluding women, but holding space is a way for men to cling to something they, even unconsciously, consider theirs. It's exhausting and hard to break into those spaces, Sheldon says. "I don't really have time to start a social movement, but I think when I'm guiding, showing up for work is quite the statement. There's emotional and mental labor that's taken to get there."

I have felt a flicker of that mental labor on so many different ridgelines. I'm often the only woman in a group of men, and when I am, I'm constantly computing a kind of attitude calculus. If I seem too soft, then I'm feeding into the idea women can't hang. Too hard and I might look like a threat. The narrative is that, as a woman in a man's world, you have to suck it up and be scrappy and tough to prove you deserve to be there. But not tougher than the dudes so you're undesirable. Sure, tell me it's all in my head, but I am always gauging group dynamics, and how I might be perceived. I can almost never let myself feel what I feel. The myth of women on the margins still holds weight, even though it was never fully true, and even though some of the best, most groundbreaking skiers of my generation have been female: Lindsey Vonn, Mikaela Shiffrin, Sarah Burke.

When my friend Mel moved to Telluride in her twenties, the ratio of men to women was reportedly twenty-two to one, and the ratio dictates the social scene. The odds are good, you get told as a new girl in a ski town, but the goods are odd. This winter, any time I rolled into town and asked which dirtbag I should talk to I was, almost without exception, pointed to a middle-aged man, like the ridge hippies and the Air Force. Those stereotypes reflect a partial reality, but they run deep, and they dictate who thinks they belong.

There are arguments and excuses in every direction, from physical to physiological: men are more risk-prone, they party harder, they're stronger so they can do the necessary physical jobs. They're better at navigating and more comfortable at being lost; they don't have biological clocks, and they have wild oats to sow. They just like it more. But the ratio also speaks to who we allow to be risky and confident with their bodies in the first place.

I spent one winter working part-time in a gear shop. When people came in to rent skis, we'd ask them to quantify their skiing ability—type one, two, or three—so we could adjust the DIN on their binding, twisting the tension higher if they were better skiers. Men chronically overstated their ability and women categorically understated theirs. You could sense it in the way they sized up the release forms, and in the questions they asked. These are the same men who will hike up something steep and sketchy with their blood in their ears the morning after they fly in from New York, or sign up for a spendy weeklong heli-skiing trip. My friend Jooles, who manages a heli-ski company, says eighty-five percent of their customers are men, even when they try to reach out to women. Confidence comes from positive reinforcement. The guys at the rental shop nod approvingly when you say you're a type three skier. Older dudes on ski patrol pick younger dudes to do control work with them. If you fit the mold, your presence is affirmed at every turn.

It *is* different to be a woman skier. It's not because we like it less, but because we're forced to be responsible in different ways.

Many of the women I know who have stuck it out in ski towns have solid jobs and life paths that aren't just pegged to skiing. They're nurses and teachers and EMTs. Their lives are a little bit more full, even if skiing is close to the center. Jackson-based mental health counselor Jennifer Sofie Gulick says she has people of all genders coming to her to try to figure out the existential questions of how to grow up on the mountains, but she's noticed women question it sooner. Her female patients often feel like they need to be responsible or closer to home even before they have kids. They tend to worry more about the physical risks of skiing. When Jooles has asked women about why they *don't* sign up to heli-ski, the two main excuses she gets are expense, and guilt about being gone for so long. She never hears the second one from men. I've seen how much the burden of responsibility falls on my female friends, even the wild children, like Rachael Burks, who I always find compulsively clearing tables at the end of the night.

I think about Rachael's path to becoming an athlete, and, like Sheldon, how hard it's been for her to claw her way to the top. To be a woman in skiing is to scrap for sponsor dollars, for opportunities, to have to pit yourself against other girls, to be the only one. You only get attention if you're the very best.

That narrow path is the opposite of the latitude we give young white men to be messy, selfish, and focused on leisure. And it starts early. As girls we're so often taught to see our bodies as a burden instead of a tool to use and break, let alone something we can leverage for pleasure. I grew up in a family that encouraged risk, regardless of gender, and I was—and still am—scrappy and tomboyish. But for all the range I was given to be independent, physical, and dirty, I still soaked in ideas about what women's bodies should look like. I wonder how differently I would absorb risk and physical pain if I had spent my adolescence and early adult years thinking about what my body could handle instead of how

I could make it disappear. In the dude-dominated world of ski-ing, where they named the bunny slope after Playboy Bunnies, I feel the burden to buck that stereotype. To be talented and responsible and kind and fun and hot, with so little room for error.

LADY SHRED

Even when there is a girl in the picture, she is so often singular. Pick your narrative. It could be Suzy Chapstick Chaffee, the ski ballet star of the early freestyle scene, or Ingrid Backstrom, the girl next door who is the touchstone of 21st century ski flicks. Consider Jenny from the film *Out Cold*, the low-budget, widely-panned movie in which Zach Galifianakis has sex with a hot tub vent. Jenny was the best friend who could keep up, and who finally got the guy. Looking back I cringe, but at the time Jenny seemed like an attainable role model. I saw her, and thought maybe I could fit that mold.

I still fight that instinct, to try to be the only one, to feel special. I can be jealous and skeptical of younger women in a way that I know is toxic. I've been told for so long that the only way to fit in as a woman is to be the only woman around. But that's not really true, and thanks in part to people like Sheldon and Rachael it's starting to change.

Early last season I called Alex Showerman, who was just back from a day of sled skiing on Vail Pass, to talk about inclusion. Showerman is a trans woman who has worked in the snow world for a long time, but before last summer, she presented as a man.

"It took me five years to come out," she says. "And a big part of that was my love for the outdoors, and my worry that I wouldn't be accepted in that community." She says she had seen from the inside how rough the elitist, judgy outdoor attitude could be and how narrow the vision of who, in her case, a snowboarder could and should be. "We make fun of anyone who isn't us. That creates such huge barriers, and if you're a traditionally underrepresented person there's so much anxiety," she says. What encouraged her to come out was the way she saw women building community, and opening space for people who looked like her. "My fear was just flat out being rejected and no longer having a home, and I think the biggest surprise to me was the embrace of the women." Alex says her only regret is not coming out sooner. Now she's consciously trying to be the person she never saw, to give other folks who might be thinking about coming out a vision of what that could look like in the ski world. She figures if she can show some closeted kid that life is good on the other side it might make it easier for them, and might make the snow world better in the process.

The best thing I've ever received from skiing is a now-years-long text chain labeled Lady Shred. It's filled up by a group of women in skiing: coaches and writers and company owners, but also teachers and bartenders who love to ski, and who make up my personal constellation of ski world friends. We glommed together initially because we felt like outsiders, but now my phone pings nearly every day and makes me feel like I'm part of something. It's a constant stream of job advice, avalanche beacon technology updates, and songs. Ski pics, dirty jokes. A place to rant and dump and brag and ask for help. It's exactly the thing I didn't find at first, when I was so spiky about being the only girl. Sheldon calls it friendtorship. We're building inclusion from the inside out.

CLOUD 9

Ski town economics

THE SKIER'S CHALET

The stairs to the Skier's Chalet are treacherous. Ice slicked, slanted, and—like the whole brown, blue-shuttered building—slightly uprooted from the foundation. It's a stark architectural contrast to the slick glass of second homes around it, but it wasn't always out of place. When Howard Awrey built the Swiss-style Chalet in 1965, it was the swankiest spot in town, the first ski-in/ski-out hotel in Aspen. Cary Grant stayed there, and so did the Kennedys.

These days, the pool is empty, save for Petey the plastic penguin slash beer funnel floating in the snow, but the Chalet is a different kind of Camelot. It's home to my generation's slopeside skids who are holding on to a legacy dating back to the '60s. It looks different now, sticky and gritty, caked with the patina of decades of parties. Currently, the old hotel lobby doubles as a game room, and triples as a ski shop, and stands in as a place to sleep if you don't mind peeing outside. It's almost always open for visitors, if your standards aren't that high. And because this is Aspen, you can often stumble into some kind of ridiculous festivities (the next day I would attend a novelty '80s style powder eights competition sponsored by a dating app).

★ ★ ★

Now, because of bigger economic forces, the Chalet's future is tenuous. The owners plan to raze the building and replace it with high-end condos. The local skiers who live there are hanging on for as long as they can, but they don't know how much longer that might be.

When I showed up at the Chalet I braved the stairs, banged on unanswered doors, then made my way around to the back of the building, where Pat Sewell was sitting on the porch in the fading light, drinking beer and looking up the 1A lift line. "Hey, darlin'," he said when I plopped down next to him. "You need one?" I did. Because this place gives me a confusing knot of feelings. Elation and angst in equal parts.

In Jackson, and in other places where I asked about the prototypical dirtbags, conversation almost always turned to the Chalet. "Have you talked to Pat and those guys?" people would ask. "They're probably the last holdouts."

I'm sure there are other holdouts, but I know the Chalet, because the residents are my friends and loose acquaintances. Their lives have braided through different parts of my ski history. I can call Pat and ask to crash because I know he's too softhearted to say no, even when the guest room is full. Over the years, the Chalet has become both a skier stronghold and a symbol for how the loose, freaky underground of Aspen—the thing that made it cool in the first place—is still hanging on.

Pat has been living in this high-life hovel since 2008, when our mutual friend John Nicoletta convinced a couple of real estate developers to let local skids live in the building until it was torn down, which, back then, was supposed to be imminent. Nicoletta, my not-so-secret college crush, moved into an old converted motel room in the Chalet first, scoring a high value home base for cheap. He pulled in Pat and Chris Tatsuno, another pro big mountain skier who co-starred with him in the ski-famous vlog "the Pat and Tats Show." They tapped JF Bruegger,

Will Cardamone, and a rotating cast of shiny dirtbags, semipro skiers and professional partiers, who made up their social scene. Fun and wild and loose, the Chalet became a destination again.

They were beating the system, paying a few hundred bucks a month to live slopeside in a fever dream of friends and fresh snow. Hints of that dream still linger. The floorboards might be jagged, and the pipes might leak in the middle of the night, but residents can bike to their jobs in town—like at the Red Onion where Pat and Casey Vandenbroek bartend a few nights a week—and they can stumble a few steps up the hill to Lift 1A, a classic chair that was the longest lift in the world when it went up in 1946. They have younger, beautiful girlfriends, and a penchant for impromptu parties. JF, who is the unofficial property manager of the Chalet—he has two rooms linked together, with a makeshift Instant Pot kitchen between—often has bacon in his pocket, and a fondue set in his backpack, so he can après wherever he goes. Once, when he broke a toe skiing, he just loaded the lift with one ski and kept going.

Now, the fantasy is on the rocks. In 2015, the Skier's Chalet sold for $22 million to a local hotel developer. The residents have been on tenterhooks since then, at the mercy of the market, unsure if each winter will be their last.

In 2019, after a long and contentious election season, Aspen residents voted to knock down the Chalet as part of a neighborhood redevelopment plan. The controversial public-private partnership included a revamped Lift 1A, and a new ski museum, along with a massive timeshare unit and a luxury hotel. It passed by twenty-six votes, and in 2020 the developers filed their plan.

That means that at some point, what's now a houseful of ski bums is going to be the high-rent, corporately-owned antithesis of everything they love. The skids in the Chalet are sitting by waiting for it to happen to them, because they don't have any power in the decision even though it will evict them from their home, and probably their community.

A lot of the guys who live there are from here—Pat's dad is the Snowmass mountain manager who met his mom when she was dancing on a bar table—but that doesn't mean they can stay. If you're Pat or Will or JF, or one of the other dudes who grew up in the valley and now work as bartenders, guides, carpenters, ski instructors, and many of the other occupations necessary to keep a town like this running, it's nearly impossible to make a living. The unromantic logistics of cost, and the squeeze of real estate in a tiny town that has historically prioritized wealthy visitors, might send them down the valley, to where housing is cheaper but the mountain is farther away, or to another, more affordable ski town—if they can find one—far away from their community.

In the thirty-something years that people like Pat and I have been alive, the ecosystem of places like Aspen has fundamentally shifted. Wealth consolidation has pressurized the middle class; housing costs have skyrocketed, thanks in large part to second home owners; the business structure of ski resorts has changed to devalue localism; and climate change, the great inequalizer, is shrinking viability.

It is achingly beautiful in this valley, intense and clear, and the town is perfectly framed by mountains. The red of the cliffs underscores the white of the snow, and the contrast feels Technicolor when the sky goes blue. But the human footprint here is less idyllic. The Roaring Fork Valley is also a dollar-hungry cesspool of designer clothes and tacky high-end mountain art that lines the walls of mostly-empty million-dollar second homes. When you leave the gondola plaza at the base of the mountain and clomp through town in your ski boots you'll pass Gucci and Prada, oxygen bars and underground social clubs.

Aspen, as a ski town, was supposed to be paradise, a resort where people could escape their everyday lives and consider higher truths. In 1894, miners pulled a 2,340-pound silver nugget out of Aspen's Smuggler's Mine, in the area's first high-dollar heyday. Back then—after the land was taken from the Utes as so many ski mountains were—it was an industry town, booming and busting. But extraction bottomed out before the economy turned to skiing. In the beginning of the 20th century, when the mines tapped out, the town's population shrank from 10,000 to 750. Aspen was dead quiet until the late 1930s and early '40s, when skiers started running a rope tow up the face of Roch Run, on what's now Aspen Mountain.

Elizabeth Paepcke arrived in the first wave of skiers. She came in 1938, hitched a ride up the back of Aspen Mountain, skied down the face, and became enamored of the place and its potential as a ski area. In 1945, she convinced her husband, Chicago-based businessman Walter, who had become wealthy by manufacturing cardboard containers, to move to Aspen and invest in the nascent ski operation. Walter worked with former 10th Mountain Division soldier Friedl Pfeifer to plan the first ski lift in town, 1A. In 1946, they founded the Aspen Ski Corporation with two other 10th Mountain veterans, Johnny Litchfield and Percy Rideout. Walter built the airport, and started laying down groundwork for Aspen's growth, trying to envision how much it might change, and what it would look like when outsiders like him started, hopefully, flooding in.

The Paepckes believed in the highbrow idea of the Good Society, in which Aspen would become a cultural center where business and political leaders could take in the natural beauty of the surroundings, and then be inspired to go home and transform their own communities. It would be a moral center for the privileged and powerful. Aspen became a ski town on the back of that idealism, which they called The Aspen Idea, and which molded the town into an influential gathering place. In 1950, the Paepckes founded the Aspen Institute for Humanistic Studies, which started as an intellectual seminar series that still pulls in presidents, artists, and academics, from Justice Ruth Bader

Ginsburg to David Byrne. A postcard-perfect ski town with an undercurrent of elite, high-minded culture.

The Paepckes' vision expanded through the '50s. They started the Aspen Music Festival, and the International Design Conference, which brought in star architects and renowned classical musicians, and made Aspen known worldwide. But that vision became distorted in the party-addled '70s. Hippies and high rollers fought over who ruled main street, and ski gangs took over the mountains, factionalizing the locals and sealing the town's image as a no-rule revelry zone. Drugs, dropouts, and draft dodgers were commonplace. Then the blowout '80s brought in celebrities, bigger hotels, bigger houses, and more excess. Aspen became synonymous with luxury boutiques and Hollywood vacations. The town doubled in size, both real estate prices and the party scene exploded, and the wholesome, intellectual ideal didn't stand still for long.

Walter Paepcke died in 1960, but Elizabeth, who was the Aspen Idea's artistic heart, lived in town until she passed away at the age of ninety-one, in 1994. She was ladylike, but prone to cracking dirty jokes and carrying a flask of whiskey, like any good ski town skid. She always shoveled herself out of snowstorms. But at the end of her life she was heartbroken by the ways Aspen had changed, and how her Eden of good ideas had become a hedonistic playground. By the late '80s she called Aspen a golden goose, and worried that it was dying, distended by the richness shoved down its throat.

Even though Elizabeth shunned the excess, the Paepcke vision of Aspen was exclusive from the beginning, full of lofty values and utopian thinking. Now, when I walk the streets of Aspen, glancing in through the art gallery glass at tacky sculptures, and other mountain house trappings I can't afford, I wonder how the Paepckes thought their vision would grow. How much was narcissistic naiveté and how much was hard-to-hold idealism?

The Aspen Idea wasn't really about Aspen as a day-to-day

community, it was about Aspen as a gathering point, a vision that ignores both the livelihood of locals, and the web of services that visitors depend on when they cruise into town for a couple of days. If you want to prioritize culture, ideas, and idyllic landscapes—the Paepcke ideals—in any kind of sustainable way you have to put structures in place to support the local community. The consolidation of wealth without any trickle-down doesn't work.

In Aspen, even the City Market is expensive. Even parking is hard. "Property in Aspen was already ridiculous in the '50s," historian Annie Coleman says. And since then, as wealth has funneled in, the cost of housing, healthcare, transportation, and everything else has skyrocketed. The town government is trying to work against that, by building affordable housing and bus routes, but billionaires are driving the line on price and city council can't keep up. Even when that local government tries to give residents some support, they're becoming desperate to find a way to create a space for people to live. They're cramming workers into substandard housing, and debating winter camping as an employee housing option. The park-and-ride just north of town has turned into a neighborhood of vans and toppered trucks, because there's no other space to live. The paradox is that the very visitors who are pushing locals out are also vital to their economy.

Pat, who works as a ski instructor and bartender, is part of that paradox. He's worried about what will happen to the Chalet, and to his life, when they get kicked out. He knows he's getting older, and he doesn't necessarily want to live in a motel room forever, even if the lift is steps away, but what are his choices beyond it?

But for now, the Chalet *is* steps from the lift, and it has snowed thirty-two inches since I drove into the Roaring Fork Valley, in that kind of all-encompassing storm that Pat's roommate Will calls the snow globe of negligence. Aspen is at the head of that valley, and as you drive toward it, the rangy ranchland pinches in, pulling storms up the valley and wringing them out. It's so cold that the snow is slow and squeaky under my skis and my toenails are going black. My shins are banged up from day after day in my boots, and I slump into my nachos at après because I'm so tired, but I can't stop skiing because I don't want to miss any bit of the goodness.

Aspen Skiing Company encompasses four mountains: Buttermilk, Highlands, Snowmass, and Aspen Mountain, which locals call Ajax. Ajax, the massif that rises straight up from town, isn't easy skiing. It's full of steep tight trees, pitchy chutes, and sneaky alleys. I follow Pat through the clear bumps, finding the soft margins, and the twin feelings of compression and float. He skis upright, his hands out wide, no wasted energy. We ski Back of Bell, the Ridge of Bell, the Dumps, all the steep, treed shots of deep snow that funnel back down to the bottom. Every lap someone new cycles into our gondola: Will and his girlfriend Jenny Harris, a second-generation ski bum; George Rodney, the former Freeride World Tour champion; ski coach Willie Volckhausen who grew up here, too, and who runs a farm down valley in the summer. There is music in the gondola, and dirty jokes. Willie rolls a tobacco-heavy spliff and passes it around. Then we are busting back out into the lung-burning cold, ready to lap again. The others flow downhill water-like, so fast and easy that I'm fighting to keep up, barely keeping them in sight. My legs are torched, but every time I follow them off the edge of a cat track, into a secret, stashy stand of trees I would have passed by if I were alone, I get a contact high. I know I'm living an illusion that people pay a lot of vacation money to chase.

Aspen is glitzy, but it's also a grind—and it's a microcosm for what's going on all over the country. When I say "living the dream" I mean finding a way to carve out a life that feels both adventurous and sustainable. When I sip my half-frozen beer on Pat's porch as the sun sinks I wonder if the ski bums of the future will all be trust-fund babies, corporate shills, and weed dealers. Did creating an industry crush the soul of a sport, or was the idea of soul always a marketing scam?

AFTER THE GOLD RUSH

If you're visiting a ski town, you might not think about the embedded economics—I don't often when I travel—but the outdoor industry in the US is an $887 billion economy. According to the Bureau of Economic Analysis, in Colorado, it's 3.3 percent of the state's GDP, and the ski industry, which is the state's largest sector, accounts for more than a third of that. Annie Coleman says the perceived goal of an outdoor trip is about connecting with something simple and wild, but the ski industry is often neither. While resorts create jobs, they can crush the culture and the landscape. The tension between locals and visitors has pulled tight through all of ski history. In 1974, Aspen doubled its season pass prices to try to keep local yahoos from skiing too much, because tourists were scared of them. Ski gangs were swarming the slopes, and ski patrol held ski jumping competitions off the roofs of on-hill restaurants. I'm sure it felt intimidating to anyone coming in. But those locals have always been a part of the draw, too. The best skiers on the mountain, the dirtbags living the high life, make it look exciting.

Pat works as a ski instructor. He's one of the pros that people coming in from New York or San Francisco want to ski with. His clients take him to absurd places all over the world, from faraway destinations like Portillo, Chile, to local ones, like Cloud Nine, the infamous champagne bar at Highlands, where—if you can afford to—you go to get lacquered in bubbles during boozy lunch. When Pat tips off his clients to the best restaurants in Aspen, or takes them out to drink with his friends, he's showing them the image of a life that feels like a vacation, the false narrative of living in a stress-free dream town. Visitors want to live like the locals, even if they don't *really* want to live like the locals. When you come in for a week, you can imagine that your life could exist between early-ups and après afternoons, because you see someone like Pat doing that. But you don't see him duck paddling under the surface, trying to keep afloat.

It's part of Pat's job to subtly sell the appeal of Aspen, and make it seem dreamy because the dream keeps people coming back to spend money. In Pitkin County, where Aspen is located, fifty-two percent of jobs are in tourism. Aspen Skiing Company is the biggest employer in the county, and the only one that employs more than 500 people, including thousands of workers during the season.

In 2018, the average wage in Pitkin County was $57,772, while the state average was just over $63,000. Average wages in Pitkin for Accommodation and Food Services, the sector that encompasses essentially any kind of tourism job, were $35,048. "Tourism has the most jobs, but they pay the least," says Rachel Lunney, Director of Northwest Colorado Council of Governments Economic Development District.

She says those discrepancies cascade out in all sorts of complicated, deceiving ways. For instance, unemployment rates are low, which is often a sign of economic vitality, but here it's a false signal of stability. Because the economy revolves around lower-paying service sector jobs, people tend to work more than one job, a statistic that isn't usually tracked. They might be employed, maybe even more than full-time, but they're struggling to earn enough to get by.

Lunney says that local wages haven't changed much in the last decade. They're barely keeping pace with inflation, and while wages have stagnated, other sources of income have flooded into the county. From 1970 to 2017, while average earnings per job grew thirty percent, per capita income grew 301 percent. In 2018, Pitkin's per capita income was $143,812, almost three times the average wage, which means that most of the earnings in the county weren't coming from wages or the local economy. Instead, Lunney says they're likely coming from investment income, dividends and retirement income, and other accruals of outside wealth. "It's really hard to track people who are making money other places but spending it here, and living here," Lunney says. And those people who make money elsewhere fundamentally change how goods and services are valued.

The growing gap between wages and wealth is a national trend. In the past decade, remote work options have skyrocketed. A 2018 survey by Global Workplace Analytics and FlexJobs found that the number of people working remotely increased 159 percent between 2005 and 2017, and the pandemic pushed that even further. People who are making money in, say, Silicon Valley, can come live in Aspen without directly needing to derive money from the local economy, which has accelerated economic disparity.

There's a metric called the Gini coefficient which measures inequality by plotting dispersion of income. A Gini coefficient of 0 indicates perfect equality while 1 means total inequality. In 2017, when the US Gini coefficient was .49, Aspen's was .65. Megan Lawson, an economist at the Bozeman-based Headwaters Institute, who studies community development in the American West, says it's an indicator for how far apart the haves and have-nots are. She uses it to try to capture the experience of people working in communities like Aspen, and to quantify why it's hard for them to keep pace when expenses like housing keep rising. The disparities are wide and getting wider. According to the Colorado Center for Law and Policy's self-sufficiency standard, a Pitkin County family of two needs an annual income of $71,274 to make ends meet. That's way above the 2021 federal poverty benchmark of $17,429 for a family of two, and the highest standard in the state. It means that in Pitkin County, a quarter of households fall under the standard. In this picture-perfect utopia one in four families is struggling to pay their bills.

It wasn't supposed to be like this. The ski bums were supposed to thrive and beat the system. Skiing was supposed to make things better. That's the heart of the myth that dates back to the days of Warren Miller sleeping in the Sun Valley parking lot. The recreation industry has long been heralded as the rural West's great economic answer to the demise of the extraction industry. Wide swaths of the West have followed the curve of mining and drilling, the boom and bust of silver and gold, or uranium and radium. You can see it in the mountain town names—Leadville, Silverton, Gold Bar—and gold panning tourist attractions. But any kind of extraction inherently can't last forever. In the mid-to-late 20th century, as mines closed and fracking consolidated oil and gas operations, recreation was supposed to create jobs in rural areas without as much environmental degradation as resource extraction.

If Aspen had only been a silver mining town it probably would have faded long ago. In her research, Lawson has found that across the West, people are moving to counties with recreation opportunities more quickly than to other places. While job growth may be faster in those towns, that growth isn't necessarily sustainable.

When I was scanning lift tickets I was told people like me were a dime a dozen. They didn't need to pay me much, or treat me well, because there was always a line of kids fresh out of college in New England who were happy to do those jobs. High desirability could keep the wages low. "You're living on vacation, what do you have to complain about," they'd tell us. But that desirability is tenuous. As inequality increases—and as expenses grow, and wages and benefits stay flat—the economics becomes untenable for many. Last winter, the marketing staff at Jackson Hole Mountain Resort were bumping chairs during Christmas week. The resort didn't have anyone to work lifts because so few people in that role could afford to live there. Alan Henceroth, the CEO of Arapahoe Basin, told me they've been struggling to find people to work ski school and food service jobs, because there's nowhere affordable for them to live. Some resorts are doing a better job than others at holding on to workers. In 2019, Aspen Skiing Company instituted a $15 minimum wage. The resort has also helped to fund public transportation, and was one of the first to offer employee housing. But there's still a big disparity between what those resorts are making, how visitors spend dollars, and how money trickles down to the workers scanning tickets on a storm day.

Today, personal jets fly into the Aspen airport, and developers have wedged mega mansions into almost every buildable space. People like Pat are the casualty. He can't exist the way his parents could.

One day after skiing, Pat and I stop into Mi Chola for après nachos. We slide into the bar in our beanies and base layers next to a bickering couple who clearly aren't from here. You can tell from their bedazzled jeans and too-clean cowboy boots. We're all bingeing on the happy hour food deals that are nice if you're visiting and necessary if you're not. Pat, who is usually effervescent, has seemed down all week, and halfway through our chips he opens up about his worries for the next winter, and the rest of his life. What happens if they can't live in the Chalet? What happens if he doesn't want to sling beers and give ski lessons forever? He wants a good life, like the one he grew up believing in, but he might not be able to do that here even though being here is the whole point.

On one hand, it's illogical to feel bad for the dudes whose dirtbag fantasy dissolved around them, but as we finish our beers I'm uncomfortable pushing on Pat's pain points. Paepcke's utopia was built on the idea of having space and time for reflection, to try to find ways to make the world more just, beautiful, and good. It's brutally ironic that the town that he loved has become one of the most economically divided places in the country.

CHAMPAGNE PROBLEMS

You can capture the excessive side of Aspen if you get a 2:00 p.m. reservation at Cloud Nine Bistro, on the shoulder of Highlands Mountain. Cloud Nine is ostensibly a cozy on-hill fondue joint, but around noon each day, the music will turn up and hundreds of bottles of Veuve Clicquot will start arriving at tables at $125 a pop. If you're there, you're not buying champagne just to drink it, you're buying to spray it, to coat the inside of the cabin and your fellow revelers with the stick of broken bubbles while the DJ spins dance music and the dudes start to take their shirts off, despite the weather. This is après to the extreme: day drinking at volume eleven, nudity, hedonism, no regard for personal property or cost. Patrons might be dipping their crab in molly, like one local told me he'd seen guests doing recently, and there's probably coke in the bathroom. Regardless, there are basically zero consequences.

By the time they shout last call, and patrol tries to sweep the mountain, the dizzy spray of unaffordable champagne is everywhere. There's a snowcat on standby to take you down if you're too drunk to ski, or if you can't find all your clothes. Then they'll do it again twice the next day (there's a noon seating, too). Cloud Nine is supposedly the best-earning restaurant, per square foot, in the country.

Cloud Nine is predicated on overindulgence and showy unlimited wealth. The tacky residue on your ski clothes is irrelevant because you can afford not to worry about your possessions. The reservation underscores the exclusiveness, but if you're hot, you can usually weasel your way in. It's decadent debauchery for debauchery's sake. Being here, bingeing in broad daylight, dripping money on the floor, is one of the clearest examples of Aspen's polarity. Cloud Nine is built for a very specific crowd, but that crowd is driving the economy here. Making somewhere like Aspen sustainable for ski bums (and really for anyone who isn't in the one percent) will mean reining in the extravagance. But for some people, that's the whole appeal. They like the idea of exclusiveness and recklessness, and they get to set the bar.

None of the normal market mechanisms work here because the whole point is that it's elite. No one *needs* a $32 million house with eleven bathrooms and an airtight altitude acclimation chamber, but need isn't a relevant factor. There are more than fifty billionaires who are known to have homes in Aspen—Wrigleys of gum fame, Walmart Waltons, and a couple of Koch brothers—and when the literal richest people in the world are sitting on top of the market, in a place already rife with elitism, the economic distortion is nearly impossible to curb.

It's not just Aspen. The same is true in nearly every major ski town. Telluride, Vail, Jackson Hole, and Sun Valley are civic symbols of the power of accrued wealth. Back in Jackson, I spent an afternoon sipping coffee at Pearl Street Bagels with Jonathan Schechter, a town council member and economist who studies inequality in ski towns. He told me that because Wyoming doesn't have personal or corporate income taxes, Jackson has become a tax haven for the wealthy. When they set up residency there, they get the double whammy of a ski house (which they can often cover with tax write-off money), and a highly reduced

tax burden. Schechter says that means the county has some of
the highest per-capita income in the country, and a massive
wealth gap. The top one percent of earners in Teton County,
Jackson Hole's home, make 233 times what those in the bottom ninety-nine percent make—the highest income disparity
in any county in the country. "Roughly eighty-five percent of
income earned by Teton County is investment income, leaving
fifteen percent for wages, and that's not much," he says. "Then
thirty-five percent, north of a third of the jobs, relate to tourism, which doesn't pay well. You hire a ton of people and you
pay them dirt, that's the economic model of tourism." There is
world-class art, charity, wildlife, and recreation in Teton County
because the wealth brought in philanthropy, but there is also
widespread poverty and homelessness.

He says that for a long time if you wanted to be in a town
like Jackson it meant sacrifices. You had to give up arts, or shopping, or your family, or really any other symbol of stability to
be there. Maybe you were poor, but you got powder mornings.
Now, people who have money, power, and flexibility can be
there without sacrificing anything. Financial investors and tech
executives can work remotely—and they do—to take advantage
of the early-morning snowstorms and empty steep chutes that
skiers once gave up any kind of monetary gain to access. That
means the tram line is crowded, the base area dive bars have
been pushed out by high-end hotels, and everything is getting
more expensive.

I heard that refrain, that it's become impossible to be a local,
almost everywhere I went. It's not just a ski town problem—the
imbalance is similar to what's happening in places like San Francisco and Seattle: desirable superstar cities flooded with huge
amounts of money. The income gap widens, and the cost of living increases, led by billionaires who can set the upper boundary.

Aspen has all the problems of any big city—unaffordability,
inequality, systemic racism—but there's also a dynamic that's

specific to ski towns, because the remote location, exclusivity, and tourism economy is part of the perceived value. One of the arguments for the expense of major cities is that there are productivity gains that come from condensing a lot of smart people into the same place. The reason New York City is so expensive is because the city creates worth as a center for commerce, but that theory doesn't hold water in ski towns. You're paying for experience alone.

But what's happening to the locals in Aspen is also what's hollowing out the middle class across America: wage flatness, consolidated wealth, governmental austerity, and an increased cost of living. I'm an older millennial, a cohort where, generationally, our parents could generally do better than their own. Now we only have a fifty-fifty shot to do as well as our folks. People my age and younger saw steep earning declines right when we were trying to get on the curve of building wealth. Stagnant income and falling wages mean that there is minimal upward intergenerational mobility—a factor that used to be consistent for ninety percent of Americans, regardless of social class. The cost of traditional tickets to financial stability, like higher education or buying a house, has been rising faster than wages. It's both an inflation problem and an emotional one. Those broad cultural and economic forces not only diminish the ski bum's prospects, they pinch dreams everywhere.

I moved to the mountains after graduating from college in 2005. When my brother graduated, four years later, in the height of the 2008 recession, there was no way he was going to commit to a low-paying, dead-end job. The economy was unstable, shaken by the subprime mortgage crisis and deregulation. He's always been the practical one, but a structural change happened in the few years between when I started making decisions about what my life was going to look like, and when he did. We watched our parents' generation's savings shrink, and

any remaining sense we might have had of meritocracy evapo-
rate. Suddenly, skiing seemed even more frivolous. The need
to grow up felt more urgent, and there was no way he could
justify being a skid.

We launched into adulthood when the ground was shifting
around us. The economy changed drastically, but so did technology, communication, and the ways we interacted with each other
and the world. As the recession hit, unemployment rates were
high and so were student debts. I got a thin window of grace
in my first few years at Beaver Creek. When I showed up I was
pretty sure I could find a job, and that the world wouldn't col-
lapse around me. But that window is barely cracked open now.
I worry that we're losing the capacity and patience for adventure. The defining feature of our age—in the face of unescapable
technology, existential terrorism, and a tanked economy—is fear.

There are reasons for that fear: the ticking clock of climate
change, the unstable economy, rampant social and racial inequality, and deep political dysfunction that makes it hard to have
faith in the future. In towns like this, that fear inspires a fight
for $15 minimum wage or a desperate scramble for new hous-
ing, or a cool acceptance during a ski lesson client's champagne
shower, knowing the perception of judgment could hurt your tip.

Every winter I lived in the mountains I had two jobs. In the
morning I'd take the bus to the ski hill, where I was paid $8.50
an hour and a free season pass to scan lift tickets and listen to
vacationers with their helmets on backward question my intel-
ligence. I'd grit my teeth and grin when some sweaty second
home owner who'd accidentally grabbed last year's pass accused
me of being an idiot when it didn't work, knowing that he'd be
gone by the end of the week, and this would still be my town.
I liked being an insider, even if I had to work for it. After the
lifts closed, every other day, I'd scrape my sweaty helmet hair up
into a bun, peel off my long underwear, hope I'd remembered

a real bra to wear under my Pazzo's T-shirt, and bust down the hill as fast as possible to clock in at the restaurant.

Because the food was decent enough to be cultish, and the Dead-inspired wall art and soundtrack made both local day drinkers and visiting dads feel comfortable, we were slammed all winter. Christmas week I would come home with wads of cash, and some nights, within the first hour I'd find myself yelling "Three hour wait, might want to go somewhere else!" as families tried to push in the door while I swung past with a hot pie.

We all dated each other, and lived together, and covered for one another, and drank and bitched and joked too much. It was the most constantly intense job I've ever had. I still have stress dreams about forgetting salads or spilling drinks. On one of those three-hour-wait nights I dropped a whole hot pizza into the lap of a windburned man from Texas who had been waiting for his pepperoni. It slid out of my hands onto his khakis, slow motion but still so fast. That night still comes back in those dreams sometimes.

The double shift was enough to cover my minimal expense: rent, food, beer, an occasional gear upgrade. I'd save up enough to float in the off-season when most of the businesses in the valley shut down. At the end of the winter I'd try to calculate how much I'd need before the rivers came up and I could start raft guiding, my other seasonal job.

My first winter of ticket scanning we were told during our orientation that the average family of four spent $30,000 a week to visit Beaver Creek—significantly more than we were making in a year, on our hourly wages. We were dependent on the visitors, and they on us, but we moved in completely different spheres.

To assert the idea that we were the true skiers, the ones who had picked the right path, we rolled our eyes at the gaper visitors, reminding ourselves that we were regularly getting the powder days they paid to fly in for. We made games of dismissing them:

what was the most awkward thing you could say to a guy who tried to grope you when you scanned his pass? Who could spot the most ridiculous way a visitor carried skis?

We were romancing the lifestyle, trying to embrace the idea that our lives were their vacations, and that we were the lucky ones because of that, caught up in our own dream of living the dream. We told ourselves that even though we were the bottom rung, we had a priceless upper hand. Sure, some dude from Greenwich is getting drunk in the pool at the Ritz but you've poached that hot tub before. Management made it clear that ours was a job for people who could come and go—who could make their decisions based not on income, but on desire.

That's not so true anymore. When I drove over Teton Pass into Jackson, I stopped for a burrito at Pica's with Malachi Artice, a young Air Forcer with red hair and a squinchy smile. He's from West Virginia originally. He followed a buddy here, fell in at the Hostel, and found ways to make life in town stick, at least for now. "My idea of going to college was getting into the Jackson Hole Air Force," he says. "I skied every day for the first four years."

Malachi is both cool-guy tough and emotionally vulnerable in a way that feels surprising in a ski town. Halfway through lunch he tells me about the waves of anxiety he gets when he thinks too hard about the future. He's been here for eight years, and he currently has a pretty good gig. He caretakes a second home, so he has a spot to live, and he works in a shop, and as a personal chef, so his income is diversified. He's been badly injured, he's had friends die, and while he's aware of how lucky he is, his way of life feels tenuous. "I never had an alternative, that safety net thing," he says.

There is a trope about ski towns that everyone there either has a second home or second job. The longer you stick around the more you realize that the people who make it look easy usually have help from somewhere: no student loans, or zero debt, or parents who will help them purchase a house when they want to. The others are treading, or lucky, or both. "A lot of people here who are living a fantasy, I can't obtain that," Malachi says. "No one is here skiing every day with a house and a family without something there to support them." He jokes about his retirement from ski bumming, and says there are almost no younger ones coming up behind him.

IF YOU CAN'T LIVE HERE YOU CAN'T LIVE HERE

I spent my first winter in the mountains sleeping in a single bed in a four-room, one-bathroom, perpetually-chilly employee housing unit caked in black mold. Our neighbors were burly snowmakers from the upper Midwest who kept vampire hours. One once punched through a glass door in a hulked-out fit of uppers-induced rage and left the pane bloody and spiderwebbed for weeks.

We'd showed up blind, not knowing where to live, or what our lives would look like, and the dorm-style apartment we lucked into was comparatively cushy. We were just out of college and my standards were low, so an extra-long twin felt normal. And down the road, in Vail, workers were being asked to double up in those single rooms, switching off bed spaces with workers on other shifts. Friends of friends crammed on couches, and into overlarge closets.

We moved out in the spring, into a thin-walled townhouse across the street from an always-lit police station. I slept on a camping pad for half the summer, until a friend working at a hotel gave me one of the mattresses the Westin was throwing away—another artifact of the ski-town barter system. We had a couch guy for most of that next winter. Elliot was only supposed to stay for a couple of weeks, after he got kicked out of a friend's parent's house, but somehow winter turned into spring and his Eggo waffles and Hamburger Helper were still in the kitchen. But we had a kitchen. And space for a couch. And the feeling that we were settling in and staying, that we weren't just a few of those flaky girls who came for a season and were gone.

Our friends settled into similar shifty situations. One drew a line with tape down the middle of the double bed she shared with a non-romantic male roommate, because they couldn't find anywhere else to live. Plenty of others lived in tents or trucks, or flitted between house-sitting gigs. The unlucky ones lived way down valley, out of the ski town circle of light. For years a group filtered through a wobbly one-story that had brown recluse spiders in the crumbling foundation, and a porch that never quite felt like it was going to hold. But it was right at the foot of the Minturn Mile, the ski trail off the backside of Vail, and it had an industrial dryer big enough to ride in, so we often ended our nights there with dryer rides, tipsy and air fluffed. Just shut the door and spin.

In Aspen, the Chalet is that same kind of wonky luxury. Living there, despite the leaks and instability, and late-night visitors, means you're in the action. But what happens when affordable housing options disappear? To learn more I track down Michael Miracle, the guy in charge of the resort's housing program. We meet for a beer in the bar at the Hotel Jerome, which has been a symbol of Aspen excess since 1889. It's been everything from a fancy hotel to a Hunter S. Thompson hangout and back again. The Paepckes owned it for a while. The founders of the first on-hill ski gang, the Bell Mountain Buckaroos, briefly lived in the basement. Now it's swanky, full of smooth leather and plush fabrics, and even though it's too early in the evening for much debauchery, there's a thrum of partying in the background. Miracle rolls his eyes at the boozing, which is par for the course. He says the contrast between skiing's gritty past and glossy future is exactly why his job is important. He's trying to figure out how Aspen can be a town that works for both the raging visitors and the locals who keep the system running. And he knows, personally, how hard it can be. Even though he has a high-up resort job, he's trying to face off the squeeze in his own life. He has two little kids, and he lives in deed-restricted housing. Even he can't afford a free-market house in town.

The prices are sky-high because much of the real estate inventory is in sprawling second homes, corporately-owned mega mansions, and condos converted into Airbnb rentals. For many of the people who own those homes—like an increasing number of businesses, or those fifty billionaires who have places in Aspen—price is irrelevant. If they want to buy a cute Victorian on Hopkins St., or a multimillion-dollar house up by Highlands they can. But you can't just blame wealth. Even if you eliminate second home buyers from the picture, there's a geographic crush that makes growth expensive. Development costs are high, and, in a narrow valley like this one, or in places hemmed in by swaths of public lands, there's not much room to build.

That's not new. There are Aspen newspaper articles from the '60s lamenting how hard it was to find a place to live, and stories looking at Aspen Ski Company's nascent plan to build affordable housing. Aspen, which started building deed-restricted housing for locals in the mid-'70s and formed the Aspen/Pitkin County Housing Authority in 1982, was the first mountain town to provide workforce housing, setting a trend that carried out across the country, and has shaped ski town living. "I cannot think of a ski area community in which housing has not become a responsibility of the local government," says Melanie Rees, a housing consultant who has worked with almost every ski town in Colorado. She says that the housing crunch happened all over ski country, and even though Aspen was ahead of the game, their residents still struggle. By the late 1990s, real estate values had soared, and the gap between free-market and deed-restricted prices had widened considerably. People like Michael Miracle aren't moving out of that deed-restricted housing because they can't afford anything else locally, and their homes haven't appreciated along with free-market housing. They're stuck. There's literally no room.

The 2019 Greater Roaring Fork Regional Housing Study, which covers the Aspen area, found that the gap between what an average household can afford and the median price of a home could grow by up to 400 percent in the next ten years. In Aspen that gap is already $1.4 million, and the study found that it costs the region $54 million a year in overspending on things like rent. Because of that gap, almost two-thirds of employees in Aspen commute from out of town to work. People are driving in from Basalt, Gypsum, and Parachute, places that don't even sound like real towns to outsiders, because they're so far away.

One solution to the housing crunch is for employers to provide short-term employee housing. Another is through long-term affordable housing, like the deed-restricted housing model that Aspen and Pitkin Country pioneered. When Pitkin County launched its affordable housing program—which included permanent deed-restricted houses for income-qualified locals, as well as seasonal worker housing—the goal was to house sixty percent of the local workers. Back then, public housing was associated with urban projects and gang violence, and no other ski town was doing anything like it. But by the mid-'90s the idea took hold, as other towns started to face the same crunch Aspen did. Today almost every ski town has a similar program, whether it's Vail's 700 rental units, or Steamboat Springs' policy of offering cash bonuses to landlords who house employees. Aspen hasn't achieved its goal to house sixty percent of its workforce. It's struggling to hit forty percent.

Michael Miracle says that there's just not enough land, even if there's money and desire. Older folks aren't transitioning out of affordable housing that was meant to be temporary, which means young people are hitting a wall. Even when the housing authority has enough money to build, they struggle with siting and consensus. "It's easy to be tough when you're sitting at the table demanding mitigation dollars, but it's hard when we say we want to put fifty units here and everyone says 'no, not here,' even the people living in employee housing," he says.

Rees says it's time to abandon the nostalgic idea of single-family homes and ample untouched public space. "We're going to have to fight NIMBYism," she says, about ski town's pervasive "not in my backyard" mentality. "People who have money don't want growth but it's necessary. We need zoning policy and density." She says we need to adapt the historic vision of what these towns should look like by building up instead of out, creating density, and avoiding new developments. "We have to change land use. We can't let nostalgia stop us from making a decent future."

It's happening all over ski country. In Mammoth, California, only two percent of long-term rental houses are vacant, which means places to live are nearly impossible to come by. In Telluride it's one percent. In Jackson, two-thirds of locals say they're struggling because of the lack of local housing. They live in their vehicles, or drive in from Idaho, braving the curves of Teton Pass on storm days. In Aspen, there was violence in a park-and-ride lot, where people were living in their cars, when one man threatened others with a hatchet.

Aspen's downtown is a pastel klatch of tiny beautiful homes, mostly old Victorians from the silver mining days. The main grid is only about six blocks square. As you get to the round-about on the west side of town, it spreads out for a bit into bigger places, mega mansions and developments, but for the most part it's hemmed in by wide swaths of public land. When the Paepckes came in the early '40s, less than 800 people lived in town, and they were able to snap up prime real estate in town including the Jerome and the Wheeler Opera House.

Rees, who moved to Crested Butte three decades ago in that wave of ski bums, says that things started to get crowded when ski towns became year-round towns, instead of just seasonal resorts. Summer recreation like mountain biking grew, the towns stopped shutting down in the off-season, and dirtbags started to settle down. At the same time, resorts were becoming real estate companies, building homes and vacation rentals on their properties. There's not so much money in affordable housing, so instead developers built high-end condos and fancy hotels—even the Chalet was originally a hotel—to cater to high-dollar visitors.

Miracle says his hardest task is changing the system without kicking people out of their homes, or completely upending funding structures. Aspen gets much of its affordable housing funding from a second home real estate transfer tax, for instance. "If you've spent twenty years moving in one direction, you can't just switch to another path," he says, about Aspen's history of wealth-catered growth. "You can't level out the system, even though the system is broken."

In 2017, Mick Ireland, the former mayor of Aspen, who has lived in the same deed-restricted house since 1997, did a study of housing stock and found that locals owned only a third of homes in Aspen, down from fifty-five percent in 2003. The attrition comes from a growing number of residences owned by corporations in addition to the ones owned by second (or third or fifth) home owners. From the numbers, it looks like Aspen is turning into an expensive ghost town with a dwindling overpriced rental market. "When you start extracting fees from the second home industry, it makes you very reluctant to turn it off," Miracle says.

Miracle says that Aspen, as a community, tends to want an impossible combination of things. It wants abundant lodging and housing, but it doesn't want to change the Victorian-scale footprint. It wants roads but not crowds. It wants a robust workforce, but not affordable apartments. Walkability, but not density. He says it comes from a desire to hang on to the old image of the town as an untouched paradise. "It's extreme NIMBYism, and Shangri-La syndrome," he says. Because of the community's unwillingness to meet demand, they're backed into a corner, trying to make it even slightly livable for locals. "Lots of people say, 'Stop trying to socially engineer your way out of it, let the market decide.' But give me any example where the market has solved the problem," he says. "We have to solve our own housing problem. We're never going to make it affordable for all income levels. It's just a matter of not letting it get too bad."

Rees says that when Airbnb started in 2008 it shrunk long-term housing options for locals. Small-scale apartments and slope-side houses became often-empty short-term rentals. Rees says the per-night cost for short-term rentals can be much higher than what long-term renters would pay, making them a more attractive option for landlords. Airbnb hosts don't pay the same taxes other lodging providers do, and there are minimal regulations, so the communities are trying to police them from behind. I can't really fault the owners, who are getting more money and less constant impact on their property than full-time rentals, but many of those owners are outsiders who have bought real estate as investment properties. Locals, both renters and potential property owners, can't compete.

It's hard to measure how many housing units are switching into short-term rentals, because there's no official metric for keeping track. Rees says short-term rentals also create under-the-table jobs, like in housekeeping and cleaning, and because that piece of the rental market isn't as heavily regulated, there aren't services built in for those workers. The law stipulates that if you're going to build a hotel in town and it's going to generate jobs, you have to build equivalent housing or pay a fee to mitigate the impact. But Airbnb doesn't have the same protections. And now towns are struggling to manage those independent, unregulated properties that are operating outside of the traditional system.

Miracle points to people like Pat as an example of who gets crunched in that market squeeze. "For the Aspen kids who grew up here, got degrees, came back here, their reality is, 'Holy shit, this is my hometown, and I've got to spend $2,500 to rent a one-bedroom, or move back into my parents' employee housing because I don't have a trust fund kicking out money.' You can't go home, the town has changed, it's an existential crisis."

LOCALS ONLY

The storm is tailing off when I meet up with Mikey Wechsler. Snow had been pounding Aspen all week, filling in the bumps and the gaps between the trees so Ajax feels less stippled than it sometimes does, and as the storm slows down it gives me a chance to tone down the powder panic and explore.

I'd heard about Mikey, who has a big smile behind a grizzled gray beard, and the barest trace of a Boston accent, before I'd even come to town. He's missed four days of lift-accessed skiing in the past twenty years. Total. And his habits are the heartbeat of the hardcore local ski scene that came along a generation before the Chalet's residents were on the hill. I first saw him at the Sundeck restaurant at the top of the gondola. He pulled his gloves off, flashing fist-sized snowflake tattoos on the back of his hands. Then I'd seen him again at the Red Onion, the oldest bar in town, where Pat bartends and he waits tables. I finally worked up the nerve to say hi, and asked him if he would ski with me. Sure, how about tomorrow, he'd said, and then, even though he was still working when I left the bar, and even though he's skied every single day since October—he blew me off for an early morning backcountry ski tour.

When I finally caught up with him and the Dogs of Bell, the

old guy ski gang he's a part of, it was midmorning, and they'd already been out for their tour. I followed Mikey, and Mary Fran, a nurse who moved to Aspen from Australia in the '90s, into the woods off of the Back of Bell mountain trying to track their lines in the pockets of untouched snow.

A lot of the young guys who grew up here ski with a particular posture, bodies curved forward into an aggressive C shape, knees tucked tight to absorb the contours of the mountain. They're fluid in their transitions, never losing speed. Mikey has that same friction-free stance—from far away I'd guess that he's twenty, even though he's three times that—and he waits patiently for Mary Fran and me, as we check our speed through the trees.

We plunge down through the snow-pummeled woods, slowing down briefly to traverse around the Red Sox shrine, because Mikey is another transplant from Massachusetts who still hasn't made it back home. We pay our respects to the ghost of Ted Williams and to the 2004 team, clicking our poles against trees as we slide through. "Figured you should see this," Mikey yells over his shoulder, before he dives downhill again.

The shrine is a couple of goggle- and jersey-covered trees that you'd miss if you didn't know where it was. There are a hundred-something similar shrines scattered in the woods around Aspen: Bob Marley's smoke shack, a memorial for dogs, a paean to Prince, usually made by locals, long ago, and tucked into subtle stands of trees, updated and added to over time. They're a small-scale hold on the local weirdness. Semi-secret, some more than others. One of those hints that Aspen is still a local's town, even if it might appear otherwise.

We sail down to the bottom of the gondola, arching turns in the ice-sheet groomers, breathing hard as we pop our skis off and get into line for another lap. Mikey's ringtone is a dog barking and it goes off when we're in the gondola, trying to defog our goggles and wipe away our snot. He makes plans to meet up with another Dog, and I ask him to explain the ski gang culture. He says it started in the '70s, as a rejection of the influx of high-dollar fanciness. It was a way for locals to bond and delineate their identity—whether they were there for the skiing or the scene. There were stylistic differences and some turf warfare. The Dogs were the second real ski gang, after the Bell Mountain Buckaroos—who Mikey says were more in it for the show than the skiing. In the early days there was some gang animosity, like ski battles and bar fights, but now it's mainly just a way to have a built-in ski crew, and to keep a sense of community and a sense of self. It's a way to hold space in a fast-changing town. Jenny Harris, a Chalet regular, who grew up here, and whose parents are Dogs, calls Mikey a powder monk, ascetic and simple. Maybe it's a little obsessive, or maybe he needs the structure that skiing gives him to survive, but the repetition makes sense. I get the value of a life rhythm and community, even though Mikey says it's harder now. "This isn't the town I fell in love with thirty-eight years ago but it's still one of the best places I know," he says.

There's a deceptive flatness to a mountain town, because the bartenders and the bazillionaires ride the same gondola car to the top. They're there for the same thing, but that doesn't mean that they can act on their desires in the same way. That's where love really jacks you up, especially the kind of long-term devotion to a pursuit and a place that can't love you back. Like so many other stories we tell about romance, the ski bum dream isn't about *being* in love, it's about *falling* in love, and the thrill and adrenaline of newness. That's where the love breaks down. The Cloud Nine visitors and the vacationers get the crushy, early-day jitters. They come for a little and leave. But if you live here it's different. Mikey's stuck it out through the dry winters and the cultural changes, to the point where the relationship might be abusive.

Mikey's ski season starts sometime in September and goes until about July. In the off-season he's on his bike, keeping his legs strong until the snow falls again. He's fanatical and dedicated, searching for snow across ranges and over passes. When Loveland and A–Basin, the high-elevation areas that open earliest, start their seasons in October, he's there on day one, grinding turns in the man-made strips of snow. Aspen Skiing Company has made marketing videos about him, and the resort, which gives out 100-day pins to skiers who ski 100 days a year, made a new one for him, when he skied 100 days ten years in a row. There's something aesthetically beautiful about that singular devotion, and how it's shaped his life. But that devotion doesn't solve his housing issues, or make the town more affordable.

There is an argument to be made that ski bumming was always meant to be temporary. That you can shun responsibility for a few years but then you should move on, and that the industry is geared for turnover because the churn of business depends on young bodies. They need new lifties who will work cheap for the benefits of ski breaks, and instructors with fresh backs to work the three-to-six-year-old ski school, where they'll be bent over crying children all day. It's novel until it's not. But when you're years deep, how do you pivot?

My last night in town I go out with the Chalet crew, to feel some of that glory that comes from sticking it out in a place like this. We are all jazzed up on fresh snow, and we sail into L'Hostaria, where one of their girlfriends works. She slides free appetizers onto our table, and zeros out the drinks on our dinner bill. We weave over to the Onion, where Casey from the Chalet is bartending, and there are beers in front of us before we even make eye contact, despite the crush of people at the bar. Mikey is working, too, hustling past tables, winking when he runs by. It feels like I'm in on the best secrets of the local underground economy, which sit unseen, just under the skin of the too-fancy town, and which help the people here get by, and even thrive in this ridiculous place.

At the end of the night we stumble back to the Chalet. I slide into my sleeping bag on the box springs of the "guest room," where skids are always welcome. I can hear the tempo of the Chalet through the walls: pipes flushing, movies running, people going up and down the stairs. The Chalet could be flattened by the time I come back next season. Maybe it's not supposed to work forever, but I want it to. In the morning I will be gone, driving out of the valley in the grainy beginning of the next storm, the road slick under my tires. I have the luxury of letting go, and of searching for somewhere else while Pat and his crew try to hold on here.

SECTION 6

A THOUSAND WORDS FOR SNOW

Climate change and the future of winter

SOUTHERN SNOW

After leaving Aspen I head south, following the Arkansas River down the Rockies into the San Luis Valley. I get a speeding ticket outside of Antonito, hopped up on gas station coffee and the kind of unbreakable early-morning midwinter light that seems to only exist in high-elevation valleys. This valley, in particular, has a certain unruly force. I have a hermit uncle in Crestone, who moved there because the energy was good, and an ex in Alamosa who tried to move out of there because the energy was bad. I roll past the UFO watchtower near Hooper, and the albino alligator farm just south in Mosca, alone on the road for the first time in weeks, feeling the buzz of new adventure again. I cross into the New Mexico high desert, still thinking about those lives that aren't quite mine, trying to keep my speed in check, trying to make it to Ski Santa Fe to surprise one of my best friends for her birthday.

While I was getting hammered by the storm track in Aspen, the mountains just south had been high and dry, barely holding on to a thin snowpack. So, instead of mourning an impossible powder day, Katie's husband planned an on-hill costume party to celebrate. At the ski basin I find a jean jacket and a headband in the trunk (rule number one for the traveling ski bum: always bring a costume) and catch them in the base for the surprise. I am not the only person skiing in jeans. Katie, who grew up in Sun Valley—her mom was one of the few female patrollers there and her dad managed the mountain—is one of the most beautiful skiers I know. Every one of her turns seems effortless and perfectly balanced, and here in New Mexico she has a group of friends who can keep up with her speed and grace. They're ex-racers and patrollers who prefer the smaller, wilder mountains. Ski Santa Fe feels like the opposite of Aspen. I never made it to Cloud Nine, but I know that the sundeck here, where we slam Frito pie and green chile dogs amidst families, is very different.

The weather is different, too, at least right now. In Aspen, troughs of low pressure trended out from the Pacific and smashed into the west side of the Continental Divide, dumping snow. Businesses instituted storm rules, the streets were lined with snowbanks, and the bottom of the sky felt close. But I got lucky there, and here, even just a day's drive away, the snow is thin and sun-beaten. The imbalance in skiing isn't just about resort management or real estate. The unfairness of these steep southwestern peaks is that, despite their perfect-featured pitches, and terrain full of aesthetic ski lines, storm tracks can often trace just north, or dry out before they make it here. And you can't ski if you don't have snow.

We follow Petra and Jack, who both grew up in the area, through the sun-crusted bumps and slick groomers. Jack used to ski patrol here, so he knows all the nooks, and he tells us about some caves we can explore. He takes us out through a shoulder of trees, then shows us where to click off our skis and drop into the underground ice tunnels. There are prayer flags and carved-out seats in the cave under the mountain, and we sing to Katie, down inside the earth. I forget, for a second, that we're missing out on snow.

★ ★ ★

Snow is alchemy, the exact right mix of cold, water, and air. You can feel the difference of man-made—the stiffness and the catch—and the unbroken crystals of newly fallen fresh. Snow is sound, too, the creaky stick of cold storms, the ball-bearings swish of slush corn or the crackle of rime ice. That thing about the thousand words for snow is right.

If we lose the ebb and flow of winter—which we soon could in places like this—we also lose storm chasing, and the barometric adrenaline of waiting for a storm. There's a risk in pinning your heart to weather, and as skiers we hang everything on snowfall. You know you're in trouble when you're constantly watching storm tracks and snow gauges, trying to predict the places where it's going to be the deepest. Letting the real-life logistics of where you're going to live, for instance, fade into the background while you focus on La Niña storm tracks, and *Farmer's Almanac* predictions.

The greatest existential threat to skiing is the winnowing of winter. The viability of these towns, and of the sport, is dependent on snowpack, which is being decimated by global warming. Depending on the emissions scenario you choose, snowfall is predicted to shrink by up to a third by the end of the century. That thin margin of weather is going to have a huge bearing on the future of skiing, and on whether or not people can keep counting on the seasons to eke out a way of life. Not just here, in the dry Southwest, but in British Columbia, where freezing levels keep creeping higher, and in New England, where almost every ski hill now depends on man-made snow. That problematic future is easy to forget in deep winters, but it's abundantly clear in shallow ones. Skiing is one of the most carbon intensive outdoor sports, and as it snows less, or rains more, it takes more energy and water to create snow.

The worst winter I lived in the mountains I volunteer ski patrolled at Arapahoe Basin, and in the early season we sidestepped the steeps of Pallavicini Face, packing down the snow with our skis so it would stick to the hills. We were trying, vainly, to hold on to some kind of base, to keep the mountain open. Mainly we were trying to hold on to our sanity, and protect everyone else's. When it doesn't snow, the land doesn't look right. A low-slung depression takes hold of the community. Everyone gets antsy. A couple of dry weeks in a ski town makes you wonder about the value of waiting for weather. Desperation sets in, and that particular season turned into a series of pray-for-snow parties, and burnt ski bonfire sacrifices to the snow gods. I did a lot of groomer skiing dressed like a hot dog to make things feel even a little bit interesting.

But every new season hinges on hope for deep powder days. There's a whole ski film industry that's built on tape cutting the highlights of last season's deep days, and delivering them to you on the threshold of winter. Even my own memories seemed to be organized by snowstorms. Did we used to talk about climate this much? Was it always this dry in December? Were we always this paranoid? It's getting drier and warmer. Can we actually keep doing this if it gets worse?

YOU DON'T NEED A
WEATHERMAN

One of my deep-day gurus, who I try to force to answer impossible questions, is Joel Gratz, a snow nerd in the purest form. He's a meteorologist as well as an obsessive skier, and in 2007 he started OpenSnow, a forecasting website, which looked at storm cycles for Colorado and has since expanded to cover much of the West. It started as a source for data-driven weather modeling for social media–savvy ski geeks, but it tapped into an obsessive fascination that many skiers have with storms. Now, he says he gets questions about bigger picture climate, instead of just weather, and how that's impacting ski days now and into the future.

SNOWTEL history and past weather models all show temperature increases of a few degrees over the past forty years, in mountain areas and beyond. Joel says the thing that makes him stressed, as a skier, is that low temperatures are increasing faster than high temperatures, which means that historically cold places are spending less time below freezing. In the past here in the high desert of New Mexico, the mountains have held on to cold temperatures even in dry years, but as that changes it means that all the climatic factors that create snow are compromised. The warmer weather has implications for snowmaking—you can't make snow if it's not cold enough—as well as snow quality and quantity. It changes the kind of precipitation we see, too. Powder comes from cold storms. Even if we get the same amount of precipitation, if temperatures are higher, more will fall as rain, especially at lower elevations, and it won't stick around. Joel also says that warmer temperatures, even with the same amount of precipitation, can increase drought because heat leads to increased evaporation. "Let's just say we have the same trends we see now, you don't need a degree in meteorology to see that seasons will get shorter on both ends," he says. "But it's hard to plan thirty years out. Even if you see the freight train coming down the track it's still crazy hard to wrap your mind around."

When he does think about modeling snowfall out into future decades the thing he thinks about—and the thing that people ask him about the most—is where skiing will still be good. "That gets to be a really interesting question," he says. "All things being equal, following higher elevation is better. Breckenridge at 10,000 feet is more insulated than Tahoe, at 6,000." He grew up in Pennsylvania, and he says some of his favorite places to ski, like Heavenly, in California, and Hokkaido, Japan, are at low elevations, which means they're at risk. But high places, like New Mexico, aren't in the clear either. He doesn't think skiing will be gone in his lifetime, but he had a kid recently, and now he's thinking a lot about what winter will look like when his son is in his sixties and where he might ski.

It is hard to track the bigger picture of climate change across a couple of seasons, but when you pull back, and look at historic change—glaciers melting, seasons shortening, and which ski areas haven't been able to open consistently—the trend lines are clear. Joel's words about variability and elevation haunted my winter. Sure, I'd lucked into a storm cycle in the high peaks of the Elk Mountains of Aspen, but in Montana, where the base of Bridger sits at 6,100 feet, it had been dry and unusually warm for January. We'd chipped through thin, rocky ice on the ridge, which usually holds deep currents of snow. In Jackson, even though we'd seen snowfall, the storms had come in hot and slushy, right on the edge of freezing. On the East Coast the season vacillated between bitter cold and balmy. Big storms followed by springlike melt-offs. The temperatures had nearly broken records on both ends.

SHRINKING SEASONS

The erosion of winter isn't just a bummer for single-focused ski bums and weather nerds. Rising temperatures and shrinking snowpack impact water supply, food security, and economic viability. Shorter, warmer winters, and precipitation that falls as rain instead of snow, screws up everything from electricity generation to fish migration. When the skiing is bad, everything is bad.

The scientific journal *Geophysical Research Letters* found that the snow season in the western US has shrunk by thirty-four days since the early 1980s. "The overall decline in snowfall has dampened profitability given the fact that industry operators have to incur significant costs in using snowmaking equipment," the study read. It's a hard economy to hold on to even when snowfall is consistent. Ski resorts have launched and faded in the lifespan of people like Mikey or Benny, who have witnessed the evolution from rope tows to mega resorts, and by the time Joel's kid is old enough to work at a ski resort the world of skiing could change even more.

A 2018 report from the Natural Resources Defense Council, and Protect Our Winters, a climate advocacy nonprofit, found that changes to winter driven by climate change were costing the ski industry approximately $1.07 billion a year. Liz Bura-kowski, a climate scientist at the University of New Hampshire who worked on the study, says that visits during the five high-est snow years in the century were 3.8 million higher than av-erage. In the lowest years, skier visits were 5.5 million lower. There's a direct correlation. People ski more and have a better time when it's colder and snowier. When it's not, the perceived value is lower. People commit to fewer lift tickets and ski setups. They buy fewer nachos and tchotchkes, and they don't stay for après. The whole industry, from shop sales to real estate, rides that wave of winter.

Resort managers historically haven't talked much about the weather because they're worried about scaring off investors. But like any unhealthy relationship, the lies and omissions eventu-ally make things harder. Some resorts, including Aspen, have been converting their energy sources, and reducing their carbon output, because they know it's important for their bottom line, but the conversion isn't happening fast enough.

Burakowski says modeling the future of winter storms isn't easy, because interconnected factors like El Niño, sea ice, or snow cover in Siberia create a complex puzzle. But despite the range of variables, there's a clear warming pattern thanks to the way carbon dioxide traps heat in the atmosphere and warms it up. "The trend toward the end of the century is to see winters that are eight to ten degrees warmer," she says. "That puts a lot of places right above the freezing level, the margin is small."

That means some resorts, especially the tiny ones that don't have the capacity to create their own winter through snowmak-ing, and which are in low elevations or southern latitudes, are going to have a hard time staying economically viable in the very near future. And as winter gets warmer, even the places that have invested in snowmaking won't be able to do much. Even if you have the equipment to make snow, if it's raining or hot you won't be able to keep it on the ground.

Burakowski is the kind of scientist who can both rationally look at the facts and the modeling, and hold the emotional side of losing winter in her head, which makes her good at talking about it in real terms—a piece of her job that's feeling increasingly urgent and important. She grew up skiing the same mountains I did, the weenies of New Hampshire, like Pat's Peak, where she now snowboards with her young children, and she's seen how much the snowpack has changed there in her riding lifetime.

She says it can be hard to make sense of it just by modeling ranges and temperature spikes, so a big part of her mission is making the science accessible through storytelling. Talking about how her mother used to ice fish in lakes that no longer freeze over, for instance. She says a big concern for the ski industry is back-to-back terrible winters, because after a few strikeouts, casual skiers start to lose their motivation. If a family of skiers skip a few seasons, their kids may move on to other sports. Suddenly you have a declining population, along with a weakening weather pattern.

The biggest imbalance of climate change, in almost any capacity, is that the burden isn't spread out fairly. The people who are most impacted by warming are often the groups least able to insulate themselves against it. As winter gets warmer and shorter, the places that struggle the most—the small ones in low, dry places, where funding is short—start to require more assistance. They need snow guns they can't afford, or water rights for snowmaking, or ways to pass the buck in years they can't open. Some ski areas will actually fare better in the face of climate change, at least for a little while, and those are categorically the ones that are already at a financial advantage, thanks to corporate cover. If the number of skiers remains the same, but the number of viable resorts decreases, somewhere like Mammoth Mountain, which sits at 9,000 feet and has a stacked fleet of snow guns, will be busier, while lower-elevation and lower-dollar operations—

say, Ski Santa Fe—won't fare as well. The snowy backbone of the country is already stippled with failed ski resorts.

I'm not immediately worried about Aspen, even though I know it's been through thin winters, but I'm afraid for places like this, and what might happen if they can't survive. Ski Santa Fe—despite the fact that it has thousand-foot-long chutes and steep, peppery tree skiing—caters to families and church groups. It reminds me of places like Cannon where I grew up skiing icy bumps, or community-owned Mt. Ashland in Oregon. It's Eastern Washington's Loup Loup or Southern Colorado's Hesperus, where you can night ski the creaky slow double chair. I'd like to see more places like that. They're still hanging on to the idea of winter, even if it comes infrequently now, but places like this could easily disappear.

MAKING WINTER

When I ski patrolled, we would head out before the public was allowed on the lifts to do our morning openings, carefully checking rope lines, and moving bamboo stakes and signage to mark hazards. We'd check the previous night's snowmaking and grooming. Those clear, cold mornings were some of my favorite times, but they could also be some of the worst skiing, full of frozen chicken heads and rocks that the snowcats dug up—we had to ski check everything no matter how chewed up it was. Maybe the first people up the chairlift only saw the natural snowfall, and only noticed the spiky Colorado peaks around them, but every bit of the in-bounds terrain had been considered and controlled. The hazards mitigated, the lawsuits avoided, the rough edges smoothed. The realistic future of skiing is a question of what counts as natural, what we try to create or maintain, and how long we can hold on to the past.

There's a halo of goodness around the outdoor industry, a sense that it engenders environmentalists and breeds people who want to protect the mountains. But just because you love skiing doesn't mean you're doing anything concrete or impactful to preserve it. Half of American ski resorts operate on government-owned Forest Service land, because of last-century ideals about public land use. That means that anything those ski areas do to gin up visitors, or improve the ski experience, impacts collective resources, be it water supply or wildlife migration. And that's before you even consider the fallacy of federally owned public land, and how the American government came to consider it public after taking it from Native American tribes. We've historically viewed attractive outdoor economies as benign, but just because we love being outside doesn't mean we're not overusing resources or damaging landscapes, and it's dangerous to assume otherwise. It's not just climate change and large-scale warming that impact our skiing experience, it's the way we use the resources, and the cascading impacts of snowmaking, transportation, trail cutting, and energy demand.

Think about snowmaking—which eighty-eight percent of resorts in the US use to keep their operations running in shallow winters. "When the snow was great it was great, but when there was no snow we wouldn't open," J.R. Murray, the general manager of Arizona Snowbowl told me a few years ago, when they were trying to figure out a snowmaking water supply. "We'd have ski seasons that were twenty days long and some where we got 400 inches. The difficulty is you can't plan. You can't hire and retain staff. So we needed snowmaking to stabilize things." It adds some crucial smoothing to the climate curve, but it's expensive and resource intensive, hard to sustain in a different way, especially because you can only make snow when it's freezing.

In 2020, according to NOAA, the last five winters were the five warmest on record and that's not likely going to stop. The Intergovernmental Panel on Climate Change says that under a higher emissions scenario (the path we're currently on), the total amount of seasonal snowfall is projected to decrease by ten to thirty percent by the end of the 21st century. That impacts far more than skiing. In New Mexico, where the Rio Grande is the main water source for many of the biggest population centers, the river often runs dry in the summer because of overuse and overallocation. Low snowfall in the mountain headwaters makes it even more precarious. It's all connected and it's crashing.

Liz Burakowski and other climate experts are trying to translate those numbers and predictions into feelings, to make us act, even when it feels overwhelming and dire. I get a deep gut ache when I think about losing snow, about the contrast between my childhood memories of snow and the gray slush of right now. I'm scared and sad and somewhat perpetually grieving. New England skiing feels almost too painful now. How could it have gotten this bad so fast?

Solastalgia is the name for the feeling of the world changing around you, when you were told it would be stable. It's the existential distress caused by climate change, and the unmoored feeling of the landscape shifting under your feet. It makes you homesick for your own life and uneasy when the weather changes. It's the deep unease of hot, snowless winters. Philosopher Glenn Albrecht, the man who coined the phrase, mashed up solace, nostalgia, and desolation, to capture that wavy feeling of loss. I feel it almost constantly these days, persistent and creeping in.

Psychologists say that the best way to deal with climate grief is to go to the places that restore you, to remind yourself of the tenacity of our connection to land. But that's extra painful when those spots that are supposed to sustain you can't hold snow anymore. Temperature-wise, Aspen could be Amarillo by the end of the century if warming continues as it has. Down here in the Sangre de Cristo Mountains, winter might be a memory soon. According to the Center for American Progress, New Mexico is the sixth-fastest-warming state in the country, and temperatures have increased by seven degrees Fahrenheit in the past fifty years. Projections indicate that the state will experience another temperature rise of up to 8.5 degrees by the end of this century.

On the hill at Ski Santa Fe, I ask Jack, who is probably a decade younger than me, if this feels normal, or if winter feels more fraught now, and he says definitely. That the conditions are overall drier, and while not every year is the same, you can see the trends. He and Petra, who have been together, and have been exploring these mountains, since high school, try to go on big backcountry expeditions as often as they can and lately some of their missions have shorted out, or they've had to pull the plug, because the snow just isn't there. They're trying to get into mountains that might not be accessible by snow ever again. Up until now, the narrative about exploring in mountains has been about first ascents and descents, but going forward we might more likely be talking about last ones.

WE'RE THE
GROWN-UPS NOW

As the climate changes, how do we ensure the future of the sport? In Aspen, and really in all of skiing, the person who has been asking these questions and prodding change is Auden Schendler, the resort's vice president of sustainability. He's in charge of Aspen's clean energy plans and political activism, and he's trying to unpack the existential threat of climate change while still running a business. I admire him a lot, because he seems to be able to balance realism with action.

Before I head out of Aspen I catch Auden at the sundeck on top of Ajax, where he is taking a conference call in the corner. A perk of his job is ski time, so after he hangs up, we head out. "Do you mind hiking a little?" he asks me, and when I say no he takes me around the edges of the ski area, and into the sun streak of Walsh's Run to hunt out the dregs of the last storm. The heat has baked the top layer of snow to crust, but he carves cleanly through the bumps, upright, legs together like an old-school mogul skier, precise. Auden is jug-eared and wide shouldered and he's a compelling combination of very smart and ski bum low-key—he can earnestly rattle off stats from IPCC reports, but he also tells me about his dudes' drinking club, and his days as a dirtbag doing fieldwork. He's been obsessed with the West since he read Jack Kerouac's *On the Road* as a kid in New Jersey. He came here for the access, he says, sweeping his arm at the stretch of the Maroon Bells in the distance, but he's found a mission that carries him through the mountains with purpose.

Auden, who has pushed for local clean energy and national climate policy, is often the loudest voice in the conversation about sustainability in skiing. He thinks about that founding concept for Aspen—the Paepckes' vision for a better world—and how he can uphold it. "The vision is that you come here and you're outside your normal life," he says in the gondola car, unpacking the Aspen Idea. "It's the idea of the good society."

That good society sounds good, but there's a gap between theory and practice. Auden's not ignoring how short the time-line is to reverse the effects of climate change and make Aspen more sustainable. He knows that the problems are both global and local. He's trying to come up with a model that will work here, but will also translate to places like Ski Santa Fe.

"We're trying to stay in business forever. That's sustainability," he says. "And to do that, you've got to solve climate, you have to treat people well, you have to manage the local housing, you have to manage the local politics." Part of Auden's job is to try to reduce environmental impacts and emissions, while still main-taining a thriving, profitable destination business. That means sacrificing certain luxuries, like lack of density, but he says it can also mean things like better transportation, and cleaner energy. The hard part is changing the culture, and breaking down de-cades of what we think ski town living should look like.

After another lap, we hop back on the gondola and pull up our goggles to talk. He says he thinks about Cloud Nine a lot, and how it's a distillation of the things he doesn't like about Aspen: the excess and the way expensiveness and waste has become the whole point. It's a metaphor for climate, too. Right now, Aspen is one of the most extreme examples of a consumption-based culture. Auden says the town is full of people who buck against sustainable changes like more efficient housing or public transportation, because they don't want to limit their lifestyles. "Mountain communities are often run by environmentalists from forty years ago whose thinking has not kept abreast of the development in their hometowns. They champion stasis over change, open space over density, and consider development evil," Auden wrote in an op-ed. "They hate crowds—even though crowds are the foundation of the entire resort economy."

He mentions Telluride as a cautionary tale. Fifteen years ago Telluride bought all the land at the mouth of their valley and turned it into open space. When they tried to freeze the physical footprint they cut themselves off from growth. Housing prices went up, workers couldn't afford to live in town, and the area started to feel ghostlike. Now, Telluride is building a boarding-house to try to compensate for the lack of affordable housing. The local government says it's necessary to house workers, but a boardinghouse isn't the kind of place you'd want to live for-ever. There's no pathway to becoming a local. Auden says that even if it's not prosaic or popular, those towns can't hold on to their images of idyllic isolation. You have to convert to renew-able energy, figure out transportation, increase density, and limit industry growth, because otherwise mountain towns will eat themselves by trying to fight change.

"Urban planning is the crux of a lot of our sustainability is-sues," Auden says. "One of the themes in mountain towns has been the corrosive challenge of NIMBYism, and the idea that to hold on to these wonderful places that feed our soul we have to freeze them in time. But if you freeze them in time you de-stroy them. There will be lines of traffic, no community, no one living in town. The only path to saving it is to accept that they will be different, and manage growth and change."

Part of Auden's vision of sustainability is the Paepcke-inspired model for creating the good society, full of creativity, joy, and nonlinear thinking. To house their workforce Aspen can build deed-restricted, dense seasonal housing in the town core, and support functional mass transit, from local commuter hubs, and from visitor destinations, like Denver. They can fund childcare, so workers can stay here with families. They need to pay decent wages. And they need to do all this while cutting their carbon footprint. He says those are hard problems, but they're hardly impossible, and they're crucial to maintaining the town's soul. "Who is the community for? It sure as hell isn't for the few rich people who can afford it. No one thinks it should just be for the rich," he says. "If it continues to be a sport for the rich, and if you lose East and West Coast snow, it all falls apart."

For Auden, the ski bum is central to any mountain town's survival. The value of the ski bum in society is a marker of a healthy Gini coefficient, and a diverse economy. "We have lost what I think is a uniquely American collectivist worldview," he says. "We need to get back to it, to make decisions that are for the good of society. You have to say, 'It might hurt me a little but I'm taking a bigger view.'"

If we want to save the ski bum, and the dug-in lifestyle that comes with it, we have to change the idea of what the ski bum is. Auden says that on some level, ski bumming is anti-citizenship. "It's the idea of, 'Leave me alone, I want to live in this illegal unit, and I'm going to eat ramen and not play by the rules,'" he says. But if you're intentionally disengaging, that means you have to accept that people who *do* engage get power, and make choices that impact others. You can't have it both ways. Ski bums have to show up, otherwise it's going to be a town full of mega mansions and the same economic divides and one-percent power that's wreaking havoc across the country.

Auden eventually breaks away for another meeting, but I carry his words with me for the rest of the winter. The thing that hits me the most, moving through these towns, is that we're the grown-ups now, my friends and me. We're the patrol directors and the paper editors and the people in charge. Pat could probably be the mayor here if he wanted to. We're not the newbie skids anymore. It's up to us to make things better.

Later in the week in New Mexico we skin up the ski hill in the evening, climbing the edge of the groomer as the sun sinks low. I want to hold every second of that night: the blue-pink streak of the sky, the Snickers in the hut at the top, the way the cold grips the breath out of my lungs on the way down. If I were Auden, I would live under a veil of constant, climate-related anxiety. But he says, despite the urgency of global warming, he thinks it is possible to hold on to this place in some form. "I hike a trail near my house outside of Basalt regularly, and I was looking at it the other day and thinking that the development is mostly in the valley bottom, most of the land is protected," he says. "We haven't fucked up the whole thing yet."

MONOPOLY MONEY

Resorts and the business of skiing

SKI THE EAST

By my second run of the day at Mad River Glen I am bleeding from the face. Not fatal, sure, but it's enough that I can see my lip swelling beyond the edge of my goggles. That morning, standing in line for the single chair, Eamon Duane asked me if I liked tree skiing. "Yeah, of course," I said. Duane, who coaches the local freeskiing team and has cut trails on the mountain of both the legal and the not-so-much variety, knows Mad River's every gully and drop. He took me to a dense zipper line he calls the 20th Hole. The trees blurred past me as I followed him through impossibly narrow gaps and windows. And, at one point, I zigged when I should have zagged and got punched in the mouth by a branch. Proof, perhaps, that I should never have written off the mountains of the East Coast.

Before I set off on my journey around western mountain towns I went back east for the holidays, hitting the gray edge of early winter in New England. From my parents' place in Boston, I borrowed a car and drove a familiar route up 93 then 89, cutting across New Hampshire and into Vermont. Up through the gap to the Mad River Valley, to one of the places where I fell in love with skiing the first time. When I went west after college, I thought I was searching for bigger things and wilder mountains, but coming here, tracking the ski trips I grew up on, I'm not so sure anymore. The scope and scale of these mountains mesh in a way that feels right to my brain. Mad River Glen has a slow single chair, a couple of careworn base lodges, and a co-op funding program. It's simple and sometimes it feels like a faded throwback. Or it would if the terrain wasn't so complex and creative. In a lot of ways it feels like my platonic ideal of a ski hill, communal, dedicated, and tough, except it doesn't snow as much as it used to (or maybe it never snowed as much here to begin with). The blue ice of the Green Mountains gleams hard.

Duane and I scour the crannies of the mountain, finding pockets of snow I never would have imagined were there, spinning laps off of the single chair until I start to feel glad the nine-and-a-half minute lift ride is so long. We hunt for slices of powder, sliding down iced-over cliff bands, hop-turning around trees. I get hung up, snagged, and twisted in places where he ducks and wiggles through. My muscles have lost the quick-twitch memory of New England tree skiing.

If you shake the snow globe and stand back, ski towns all look the same, but there's a specific texture to all those places: the chippy red rock under Aspen, the wind crunch in Summit County and the steel chalk of Big Sky. It's the way the granitic fall line of Jackson feels harsh on every level, like you can never quite catch your breath, and how Alta always seems sparkly, as if there's a fine spray of silver in the air, catching the light, refracting. Certain valleys trap storms tight and wring them out, and each ski hill has its signature. It shows up in the way Whistler locals ski with squeegees in their pockets, and how people from New Hampshire use their edges hard.

A lot of skiers I know are obsessed with that intangible idea of soul. Mad River Glen, for instance, is soulful. It's a ski hill, not a ski resort. From the dirt parking lot you just clomp your way over to the single chair. You can probably leave your duffle under a chair in the lodge and expect the crowd to keep an eye on it. There's no base village, no shopping aside from one little sticker shop. Nothing to do but ski, and maybe get a soup at General Stark's Pub. When we say soul it sounds like the bull-wheel creak of slow double chairs. The way you can hear your edges cut through snow if it's quiet enough around. The way some people call going to the mountains going to church. But as a few big corporations take over resort management, skiing's soul is in jeopardy. Mad River Glen is a co-op, it's hanging on in part because it's so unique, but a lot of other places aren't as lucky.

OUTLAWS AND
OVERLORDS

The move toward a corporately controlled cookie-cutter future is a specter that currently haunts skiing, and it's scraping the sport's sense of place raw. In Montana, I'd met Riley Lemm at his job in the Big Sky Sports tune shop on the bottom floor of Big Sky's mountain mall. He'd come in for his shift in a bucket hat and button-down, and before he got to work behind the Wintersteiger machine he headed to the shower in the shop building. He relies on the bathroom access because he lives in a tent on some unsold property in a fancy development nearby. Riley is college-aged, with an open earnest face and a sly intensity. He approaches skiing with a studiousness he seems to bring to everything, including his not-quite-legal living situation. He didn't get a spot in employee housing, and he can't personally justify driving up to the mountain from Bozeman every day (an hour if the roads are dry, maybe forever if they're not), but he wants to be here, and he didn't see any other realistic way to make it work. So this fall he and a friend found a vacant lot that seemed to be uninhabited, set up a wall tent, built a stove platform, and moved in to squat. They play chess huddled over the stove, and build jumps on the back slope of the property on nights when the temperatures are mild. Riley says they're getting pretty good at cooking on the woodstove, but still, they're technically living illegally.

Riley is smart enough that he looked up the Montana laws to make sure he had an argument to back up his trespassing. His wide brown eyes get bigger and his voice speeds up as he tells me about the factors that brought him there, to camping in January in Montana. He thinks Big Sky is intentionally recruiting workers who don't like to ski, so they're not trying to get shift breaks. They're building hotels, but not housing, and the town is beyond its limits on water and space.

It's killing the soul skiers like him, he says, because not a lot of people are going to move into a tent in Mon-fucking-tana when the temps drop below zero and the only glory is in the empty early morning on the hill. He says his manager at the shop has tried to put pressure on the resort to build more housing. "He knew it was an issue, but it wasn't until it was us, his employees, who didn't have a place to sleep, that he actually acted on it," Riley says. "It doesn't get attention till it gets desperate, but we're desperate."

Riley says that from his perspective, the lack of housing encapsulates everything that's wrong with the ski world, and the way that it has shifted its center away from skiing. The resort management's focus is on short-term gains, he believes, which will eventually undercut their ability to keep the resort running. The imbalance comes from consolidation at the top of the resort ownership structure, which is changing the very makeup of ski towns.

Big Sky is a part of Alterra Resorts, one of the biggest resort groups in the country. In the past few years, a majority of big ski resorts have been bought up, or tied into two major resort groups, either Alterra, which was founded in 2018 and quickly started snapping up resorts like Mammoth and Deer Valley, or Vail Resorts, which owns nearly half of the North American ski resort market.

★ ★ ★

They've effectively created a duopoly for ski resorts in the US and many of the large Canadian ones, too. That business move is shaping pricing, access, culture, and crowds. They both offer passes—Alterra's Ikon Pass, and Vail's Epic Pass—that grant access to a large number of mountains. They say that it brings stability, cost sharing, and uniform experiences, but the conglomeration also means resorts and skiers who *aren't* part of the consolidation are being cut off. Vail and Alterra can set a line for lift ticket and season pass prices because there's no real chance of a competitor disrupting them. Other places can only lower ticket prices so much, and if they're not part of the big conglomerates they often don't have the collective pockets to fund big projects, or keep up with changing expectations. Somewhere like Mad River Glen, which has branded itself on its independence, can still hang, but other small ski areas are being crushed by the duopoly.

The conglomerates have set season pass prices comparatively low. As a consumer, if you are able to throw down money in the fall for a multi-resort pass, and then ski frequently across a wide geographic range, you're getting a really good deal. Say you're like my friend Lyndsay, who lives in Salt Lake City, has a van, and can travel easily—you're lucking out, ski passes have never been cheaper. But if you're a casual few-times-a-year skier, prices have skyrocketed. In 2018, the average US weekend window lift ticket price was $122.30. That's thirty times greater than the $4.18 it was in 1965. Over the same half century, US disposable family income grew slightly less than threefold. That means the lift ticket price grew ten times faster than people's ability to afford them.

Those consolidated passes push skiers to certain mountains, and target a particular kind of skier, the mobile, high-dollar visitor. Skier numbers are flat to slumping, especially as boomers stop skiing—visits for young people, and particularly young women, are down. And even though season passes are comparatively affordable, other associated costs, like gear and transportation, are up. "The US ski industry is facing increasing prices, paid by a declining number of customers," analysts wrote in the *2018 International Report on Snow and Mountain Tourism*.

Vail Resorts, which started this round of resort consolidation, has carefully cataloged and tracked its users' data, and ascertained that the most value comes from visitors willing to travel. Those who are just starting out, or who want to commit to their home mountain, as well as skiers who can't or don't want to travel to ski, don't benefit.

This bifurcation divides the haves and have-not-as-muches. It means crowding at certain resorts and emptiness at others. Almost everywhere I went this winter, local skiers complained about consolidation and crowds. In Montana, Big Sky locals griped about the ways Ikon skiers crushed the local dirtbag culture. Air Forcers in Jackson were pissed about crowds in the tram line, and Snowbird skiers faced previously-unheard-of canyon traffic jams bad enough to keep them in Salt Lake City. Arapahoe Basin, my old home mountain, dropped out of the Epic Pass, after a multi-decade partnership with Vail Resorts. The Basin was briefly independent, and then decided to join the Ikon Pass, because, from a resort management standpoint, there is pressure to be in one of the groups or be left behind.

When those resort consolidations happen, local quality of life doesn't necessarily improve. In Aspen's Pitkin County, and nearby Summit County (home to two of the Vail Resorts, as well as Copper Mountain and A-Basin, which are part of Alterra's Ikon Pass), wages have stayed flat over the past ten years. There *are* jobs, but they're not necessarily the kinds of jobs that make it easy to live in those towns. That's always been true—the liftie lifestyle has never not been tenuous—but it's getting even more stratified. Skiers like Riley, who work hourly jobs, are just barely able to scrimp by, and there are few ways to move up the ladder. If you're not as scrappy as he is, or if you don't care enough to commit to semi-illegal vagabonding in the name of powder days, it might not feel worth it.

Resort consolidation is also eliminating pathways to becoming an obsessive skier. I worry about monopolies sucking all the secrets and personality out of skiing, but I also worry that we might be spiraling toward the end of skiing for people who can't afford to roll into a place like Aspen. And if we do lose that, skiing becomes a faceless show of wealth, soulless.

I know that because Beaver Creek, where I started my ski bumming stint thanks to that invitation from Ivan, is the antithesis of subtle and soulful. It's a mountainside expanse of wide-hipped mega cabins, often empty except for holiday weeks. Heated driveways suck the chill out of the air, and the pools are perpetually warm, even if no one is in them. In the village you ride escalators to the lift, past the fur shops and fancied-out Starbucks. It feels like you're in a high-end mall that happens to be at the base of a mountain. Free cookies are served on silver trays every afternoon. Looking back, I say I would have gone somewhere more rugged and wild if I had thought about it more clearly, but I wasn't smart enough to know the difference then. The Beav was one of the last major resorts to be built; it opened in 1980. It is full of the treacly bullshit that came from early-'80s aspirational capitalism, but it also has beauty and toughness if you know where to look. Grouse Mountain, the steep peak at the back of the area, is pitched like a roller coaster and stippled with chest-high moguls. The only groomed run off the lift is the Birds of Prey downhill course, where they run a World Cup race each December. Every early season, in preparation for the event, the race crew pumps the snow full of water so the course will run fast. That icy base will stay hard blue for the rest of the winter. Unless you get it first run after a dump of snow you can barely power slide your edges in, but if you catch it early on one of those fresh mornings, right after patrol drops the rope, you get a perfect vertical zero G float, the kind where time slows down.

Whenever I go back to Beaver Creek, I get a complicated kind of déjà vu. The first bus ride back up the mountain always feels like getting the wind knocked out of me. I am rocketed back to being twenty-one. My body starts to go through the motions of skiing those specific runs. I feel something akin to grief, a longing linked to muscle memory. My personal history traces the mountain's dips and rolls: here is where I broke my thumb trying to ski backward, here is where Jono told me that "White Wedding" is the perfect song to have in your head when you ski bumps. This is where I took my first real powder run, wallowing on my skinny skis. This is where I saw a man die when he hit a tree, and where my ex-boyfriend took his body down in a ski patrol sled. A portion of my personal history is mapped onto those mountains. When I'm there I feel like I'm floating through my past, half in, half out.

This year was no different. When I stuck my head into the patrol shack, on an extra-cold February morning, there were some familiar faces, which makes it even easier to blur the line between then and now. The folks who are still around somehow look the same, just a little grayer and balder and more sun scratched. Maybe they have a few new injuries, but I do, too. There are rookie patrollers, and young strangers, but there are plenty of people I used to love, and still do. When I'm back I'm constantly scanning crowds for familiar faces, trying to pretend I'm still in. It aches. I know I'll never quite be that person again.

I end up crashing with a patroller I used to have a crush on and I'm sort of surprised to see that in the time I've been gone he's grown up. He has fresh vegetables and clean sheets. In the morning, when I leave long after he's gone to file the avalanche forecast, I walk out onto the sugar-dusted sidewalk to find that he's scraped the snow off my car. We have softened and changed, grown up. And even though it's good for both of us I get a knocking ache of nostalgia I can't shake for the rest of the day, a feeling like the raw real soulful parts of my life are behind me. They look better in the rearview.

CORPORATE CULTURE

This idea of group resorts and discount passes isn't new, and it hasn't gone particularly well in the past. In the mid-'90s, resort developer Les Otten, who bought Sunday River in 1980 and made it huge, went on a ski area buying spree. In 1994 he bought Attitash, in New Hampshire, then Sugarbush in Vermont. Over the next three years, his heavily-leveraged business, American Skiing Company, bought twelve resorts. By 1998, ASC owned a quarter of the New England ski business, and its reach extended to Heavenly in California. Otten's vision was to expand like he did at Sunday River: to turn those resorts into real estate businesses with a side of skiing. I got my first ASC pass in 2001, my freshman year of college in Maine, so I could road trip across New England to ski. By then, ASC was already on the downswing, losing value in the stock market, and starting to shed resorts. They sold off Sugarbush before I could use my pass there, already overleveraged and past the tipping point. I spent my lifeguarding dollars on a fantasy that was already starting to fail.

Skiing has always been a business with thin margins, a gamble on seasonal risk, and Otten's overly-ambitious vision crumbled on that chance. According to Ski Area Management, "We know what happened to those 1990s conglomerates: heavily leveraged real estate development and resort purchases sunk ASC." But the Otten vision of resort consolidation still lives, and resort groups are still trying to turn skiing into an all-seasons, all-hours-of-the-day "experience."

I moved to Avon, where both Beaver Creek and Vail headquarters were based, just as Vail was about to move out. In the winter of 2006, as I scanned lift tickets at Beaver Creek, Rob Katz became the CEO of Vail Resorts, effectively my boss. That year was an inflection point, not just in my own life, which shifted into dirtbag mode that season, but also in the way large-scale resorts began operating. Under Katz's leadership, Vail started buying up resorts beyond the ones it already owned: Keystone, Breckenridge, and Beaver Creek.

The next spring, Katz uprooted the company headquarters, which had been just across the Eagle River from my shoddy employee housing, to suburban Broomfield, northwest of Denver. "It's got some very nice mountain views," he reportedly said, unironically. The move changed the corporate culture and extracted management-level positions from the mountains, creating a community leadership drain. Before that, resort workers could move up through the ranks, and build sustainable long-term careers in the mountains. By moving those higher-level management jobs, the resort groups shifted centralized power away from the local economy. When Alterra started, in 2018, its headquarters were built in the suburbs north of Denver, too.

There are some upsides of consolidation. In theory, the benefits of a multi-mountain group are things like stability and consolidated savings. The ski industry is a fragile business, and any financial growth comes from trying to reach a static, or potentially declining, population. When you purchase a pass early in the year, the resort gets financial commitment ahead of the season. They can allocate funds to where they're needed, like snowmaking programs, and invest in expensive upgrades, like new chairlifts. Now, Vail and Alterra can ride out that instability because they're geographically diversified. Since they have resorts across the country, they can spread out the risk. Say it doesn't snow in Tahoe, but hammers in Colorado? Vail is still making money, while small resorts in the Sierra Nevada have to weather the bad season, which are becoming increasingly common due to climate change. A big snow year makes a big difference for any resort, especially if it comes in before Christmas, when people are buying gear and committing to vacations. Independent places aren't as financially insulated to sit through the lack of storms.

"Smaller ski areas just can't compete," says ski resort consultant Tom Lithgow, who is the chairman of the National Ski Area Associations Economic Analysis Committee. "We have a crisis right now, mainly around investment in snowmaking and replacing ski lifts, and we don't talk about it. Most lifts are over thirty years old, this stuff isn't made to last forever, and if you're replacing a detachable quad you're looking at $3 to $4 million. Somewhere like Pat's Peak, in New Hampshire, can't buy a new lift." Smaller independent resorts fall by the wayside because they can't afford infrastructure upgrades, which means they can't stay relevant. And as those smaller resorts struggle and close up shop, that also means that families who can't afford a trip to Aspen, or Breckenridge or Stowe, won't have a place to ski.

Lithgow lives in New England; he's seen the ski industry there peak and fizzle over the years, and he says it's getting more corporate, especially as the founding generation, from the 10th Mountain folks on, ages out and sells. Resorts consolidate, or are bought up by venture capital groups, who care more about profit than powder. He says we're potentially at a crisis point, because as ski areas get older and need maintenance, and as mountains need more snowmaking to account for climate change, it's hard to make gains. Infrastructure is expensive, and even non-sexy things like new snow guns require capital. That's where the big guys have power, and why places like Pat's Peak are on the edge of precarity.

The Vail model also encompasses the real estate, restaurants, and retail stores in the base village—a Disney World model, but for skiing. Beaver Creek Sports is owned by, surprise, Beaver Creek, which means they're really owned by Vail Resorts. Almost every business in the mountain's base village is now affiliated with the resort conglomerate. Depending on how you look at it, it works. From a financial side, Rob Katz is very good at what he does. Vail, which is publicly traded, has seen tenfold growth in its shares. The company IPO-ed in 1997 at $22 a share, and in January of 2019, despite less than one percent growth in skier numbers over the past two decades, it was trading at $233. Even in the face of the pandemic, when skier visits were down, Vail stock was ascendant. Because it's publicly traded, the company is responsible to those shareholders. "People don't like to talk about who owns these things," says Jamie Schectman, who founded the nonprofit Mountain Riders Alliance, which supports small ski areas. "But Vail is a publicly traded company, they try to make as much money as possible. Alterra's majority owner, KSL, is a private equity firm. The ownership group is expectant of big returns, and that affects the operating model."

That financial targeting feels scary to me, because it impacts both the sustainability of ski bum living, and the return rate of beginner skiers. In addition to the move to Broomfield, under Katz's reign, Vail has underpaid and undervalued employees (Katz has publicly said that ski instructors are a dime a dozen), kept wages low, failed to build promised employee housing, and homogenized ski areas. And when Vail makes changes, to anything from pass prices to wages, they change the baseline for the rest of the industry, which has to hustle to keep up.

Katz has publicly called the Epic Pass model Walmart pricing, and the world's largest ski resort company's business strategy isn't dissimilar from the world's biggest retailer. Once you're there, the resort makes money on lodging, parking, ski rental, equipment sales, clothing, lessons, food, and beverages. In the same way that Walmart came into small towns and undercut smaller shops, and the way Amazon has made it nearly impossible for local businesses to compete on price or variety. The mega resorts are killing the soul of skiing with monopolized ease.

What you're losing when you lose the dirtbags and family-run mountains is diversity, depth, and a sense of place—that soul. Corporate ski resorts are businesses, not culture generators, but you need to sustain some semblance of individuality to keep people coming back. Localism dies in the face of optimization. Rows of T-shirt shops and generic restaurants are mind-numbing and that facelessness feels nightmarish to me. What I like most about showing up in a new town is trying to find the weird parts. I want to be there with a local who knows her way around. I don't want to go to an eerily familiar McMountain. I want Riley to geek out on telling me where he skied yesterday, and let me in on the secrets.

LOVE, ACTUALLY

I'm defensive because I still feel indebted to the scrappy New England ski hills where I first felt the rush of gravity: Pat's Peak, Nashoba, and Cannon—the kind of low-key places that could accommodate school ski buses, day-tripping families, and duffle bags shoved under the cafeteria tables. We'd stash PB&J and peanut butter M&Ms in our Columbia Bugaboo pockets, and sometimes find the wrappers there the next season, if we hadn't outgrown the jackets. In the fall my family would go to the basement of Ski Market, a dank warehouse of pre-worn gear, to switch out last year's equipment for bigger boots and longer skis.

How do you become an infatuated, lifelong skier if you can't even start? Like any other kind of love, there needs to be tension and buildup. Some of the most fanatical skiers I know are the ones who learned to ski on the edgeless blue ice of New England, the Midwest, and the mid-Atlantic. Jeff Leger, Doug Coombs, and I all started skiing at Nashoba Valley, in eastern Massachusetts, which has a whopping 200 vertical feet of elevation gain, and so many other dedicated skiers I know started out at mom-and-pop resorts, memorizing every line at Roundtop or Whaleback or Blanford, as they grew into the infatuation. They were already hooked by the time they headed for Jackson or Squaw. The reason Beaver Creek still felt special to me, despite its glitz, is because I was a goner long before I got there.

New England skiing is all frozen nose hairs and ice cube fingers. When I go back to the scrubby East Coast hills I grew up on, it's clear that the mountains there are different, and so are the ways you move through them, pinchy and turning tight. There's hard-earned grace in the unforgiving zipper lines, and the way you have to power slide over the rime-covered cliff bands of Mad River Glen, tuned into the subtleties of the color of ice. I got soft the second I moved away. There are fewer places in the thickly treed mountains of the Greens and Whites where you can just let physics take over, and even then you have to be a little brave or a little stupid. Trying to tail Eamon through the Mad River trees it was clear he was going easy on me, waiting up.

The little hills reel us in. When you ask people why and where they started to love skiing, it's often because of church ski trips from Kansas to Wolf Creek, Colorado; soggy, Seattle-area days at Snoqualmie Pass; or lit-up Wachusett nights. In places like Big Sky, many of the displaced skiers who have moved in from smaller places in search of bigger mountains and broader ranges still wear their roots like a badge: the Loon Parks hoodie or the hand-me-down Mt. Hood Ski Bowl hat. I know people who have moved back to New England or West Virginia, because of that. They'll give up terrain for soul.

I have a physical, full-body response when I go back to Beaver Creek, like I'm regaining a piece of myself. Mikey says love whenever he talks about the mountains, so does Marcus, and I can hear it in myself. I grieve the loss of localism. I want a favorite run at a favorite hill, the feeling of rolling up to the mountain alone but always finding someone to ski with.

There are mega resorts in New England, too—Vail bought Stowe in 2017, and then continued to buy in the area—and small scrappy ones out west, like Montana's Discovery Basin or Arizona Ski Bowl. We were shaped in those less prosaic places, and those places have to continue to exist to keep those pipelines flowing, and to give the ski world some semblance of equity. People like Jamie Schectman are fighting to protect those local gems. In 2010 he realized smaller ski areas were dying off. "Back in the '80s there were 730, now we're down in the 400s," he says. MRA has worked on stabilizing little ski hills, but even if you're not trying to turn a big profit, it's a hard business. He's worked with Mount Abram in Maine, a modest family hill where day tickets range between $29 and $49. Mount Abram tried to keep everything cheap, but with overhead and salary costs their season pass was only seventy dollars less than nearby Sunday River's Ikon Pass. He says it's impossible to achieve economies of scale.

"That's a problem because those small resorts are gateway drugs," he says. "You have to walk before you run. You don't start going to Vail. You go somewhere with a lower price point, that's not intimidating, where everyone knows your name. When the big resorts say it's $900 to go to 18 different resorts, that's good and fine for the person who is already hooked, but it's a short-sighted view on the industry. Right now, skier visitation is flat, old people are exiting the sport, and you're pricing out a big piece of society."

By trying to let the algorithms make it frictionless, the ski industry ignored the best part, the sense of adventure. Big resorts, with their mechanical, bloodless view of the sport, have a narrow vision of who a skier may be and what they love. Studies have shown that people are happier when they don't have as many choices. We don't always want to be optimizing. And those huge industry investments haven't particularly paid off. Even though resorts invested billions of dollars in the nineties, skier days did not grow significantly.

Like Tom Lithgow says, we didn't build these towns, much less these chairlifts, to last forever, and it sometimes feels like they've grown beyond viability. He's a suit, but he still gets it. "I kind of think the evolution of the industry and the dynamics within the country, economically and technologically have diffused the glory of the ski bum," he says. But I want the kind of glory you find in the hand-cut glades of Mad River Glen.

THE DARK AND THE LIGHT

Risk and reward

LAND OF BROKEN TOYS

The resort biggies know that you can't just slap a ski hill anywhere. You need the alchemy of slope angle, snowiness, and space: a ridge that stops storms, and a pitch that makes your adrenaline pump, the steeper the better. You can see it most clearly from the air; the fluted fins of big mountains make me obsessive whenever I fly. I trace the possibility of skiing on the window with my finger any time I'm coming into Seattle, skimming the sides of the Cascade volcanoes, or flying down the rib of the Front Range headed for Denver.

You can find that collision of geology and weather anywhere there's serious skiing, from the Alps to the Andes, and those topographic factors are highly pronounced in Utah's Wasatch Front, where lake effect storms and dry high-altitude snow slam up the center of the rocky canyons just east of the Great Salt Lake. Salt Lake City sits in the bowl of a valley that stretches out toward Reno, and the mountains scrape up in contrast, neck-crampingly steep and filled with vertical ski lines, skinny couloirs, and steep aprons pinched by cliffs. They're most concentrated in the parallel gouges of Big and Little Cottonwood Canyons, which look like a bony knuckled fist from the air.

Zero in on the snowy steeps at the top of Little Cottonwood, and maybe you can see me following Spencer Harkins through the grit of Alta's High Traverse, hands in front, knees to chest, trying to absorb the high-speed wobble of the compressions and ruts, heart in my throat. The High T transects the crown of the ridge, but you have to hold speed to make it across the snow field. Spencer airs effortlessly over a chunk of crumbly quartzite, kicked up by the last skier to come through. I swerve behind him, back seat, trying to maneuver fast enough to avoid digging a ski tip. On the far side of the face we drop through a narrow gap in the trees, swivel across a little set of bumps, and ride one last leg of the traverse out to High Rustler. We stop, finally, on a ridge above a narrow chute. Little Cottonwood Canyon spreads out in front of us. Across the road, Mount Superior is catching shadows in the gold and rosy light. Salt Lake City, down in the valley, is under a veil of smog, as it often is, but it feels impossibly far away from here in the crown of the canyon.

There's a tricky must-make move to enter the chute. It's short but committing, and after Spencer slides through, pointing his skis to gain speed, then disappearing, I have to tamp down the fear of falling and follow him. He's let me in on the secret—I said I wanted this—and now I have to see it through, even though I'm scared. The dream can flip to a nightmare fast.

Earlier that day, my first morning at Alta, I'd found Spencer at the bottom of the Collins chairlift, in his go-to look of jeans, a hat, and sunglasses, his long blond hair flowing out behind him. He has East Coast roots like me—he moved out here from Massachusetts, too—but now he's a fixture in the early morning Little Cottonwood lineup. His job and his girlfriend and his community are all here. He was standing in line with Leo Ahrens, a pro skier who grew up here, and has been skiing sneak lines in the Wasatch since he was tiny. "We're going to ski High Greeley," Spencer said. "You should come." And because I'm not smart enough to say no to anything, and I have an overeager hunger to be included, I slide onto the chair next to them. We ride up Collins, and start sidestepping right off the top, heading toward the crenellated ridge of Greeley Bowl. Alta is a zigzaggy, off-camber mountain, full of sneaky sidesteps and tough-to-navigate traverses, so I try to follow close behind.

On the hike I'm immediately out of breath. Leo is hungover which helps the hiking pace a little, but it's still painful to keep up. The traverse is barely a ski wide, so in the narrowest part we hold on to a rope that ski patrol has bolted into the cliff. Once we slide beyond the rope we pop our skis off and boot pack to the top of the ridge, climbing through a gap in the spiny piton of rocks.

My sense of self is so often shackled to the things my body lets me do, and it's all there on the ridge: the frustration at my fast breathing and slow body; the fear of not being good enough; the slammy shin bang I get when I chase them out across the traverse, trying to hold on; and the visceral hit of relief and beauty when I step through the gap, gasping. Sometimes the good parts and the hard ones are so close together—like the inhale and exhale of the same breath.

I wonder if other people's nerve endings work differently than mine. Maybe I didn't cut it as a skid because I wasn't addicted enough. Being comfortable with risk, and having a high tolerance for instability, is another ski bum commonality, as is the ability to throw yourself into a questionable situation while negating or ignoring the consequences.

There are mental and physiological differences between people who are geared toward risk and those who are more cautious. I like the jag of adrenaline, but I don't like the paralyzing, full-body fear that often comes with it. Skiing is better and easier if you pick up speed, and the best skiers look effortless, they know when to shift their weight, and when to pressure their edges. But good skiing is more than being comfortable with speed. Some people's brains don't really fire until the consequences are high. I have friends who are guides and athletes who don't have any physical anxiety until they're navigating in a whiteout, or knuckling down on a ridge. Sheldon says her brain works best in moments of high consequence. Adrenaline wakes her up and lights up parts of her brain that don't respond in less stressful situations. People like her have quiet amygdalae, and they thrive on precarity. Couple that with antisocial tendencies and you have someone perfectly suited to this lifestyle—but the risk factors persist.

You can end your whole season in a misjudged landing, and your entire ski life in a fall. Almost every serious skier I know has blown a joint or cracked a bone. ACL surgery is almost a rite of passage. I watch my friends grow up, get older, and start saying prayers for the ligaments of their knees. They self-medicate, and stretch in the parking lot to try to dull the ache of old injuries. "Did I choose right or did I get lucky?" is a question that has started to course through my brain any time I return from the backcountry.

The risk is ever present, but one of the core false narratives of the ski bum story is that you're healthy and fearless and hot forever. That your body can withstand the seasonal pounding of banging bumps and bumping lines, and that you can be stoic in the face of broken bones, or busted relationships. Always able to joke. Always up for another. Three weeks into our ski bum stint, Katie broke her wrist in a losing battle between her ski boots and some icy stairs. She had to be hauled off the mountain by ski patrol, humiliated, but she reclaimed her shiny new-girl status by doing one-armed keg stands at a ski patrol party that evening, fearless and game.

At the top, Leo shakes his headache off, and finds an impossible-looking line through the steep cliffs, rolling easily over his ski tips, floating his turns. His history here is so long that the memory of every chute and cliff band must be burned into his brain. It doesn't come as easy for me. I am scared of his line's relief, so I ski around to a wider slot nearby, not willing to launch myself off cliffs I can't see, even if that's what we came up here to do. At some point, partway down the ridge, where the slope mellows, I stop being fearful of the unknowns and let myself open it up.

The miners who first moved into Little Cottonwood called the narrow pinch of the canyon just before the base of what is now Alta Hellgate. The canyon is intimidating and steep, and they supposedly said that nothing good ever happened past the choke point. Alta's steepness is iconic. The road through the canyon is so prone to avalanches that ski patrollers started bombing the mountain preemptively after big storms, birthing modern American avalanche control. Alta's powder skiing culture dates back to the '30s, and it's now home to a thriving pro athlete scene. You can feel the pressure of the past and the future here. The tiny ski village at the top of Little Cottonwood is a bubble of highly talented, hyperfocused skiers. Many people who work here live in the lodges and hotels where they're housed and fed. They barely leave the canyon all winter, because many of the things you might want as a skier are right here. But the partying, the altitude, and the topographic intensity can wear thin, too, especially if you can't get out.

Alta is sparkly and grand, the light here is somehow always golden, but the mountain can be harsh. People die in this canyon, from avalanches and accidents, and from their own infliction. Suicide rates are astronomically high in ski towns—Aspen alone has three times the national average of completed suicides—and so are rates of substance abuse. It is hard to push it forever, and it's particularly hard in a place that glorifies risk.

I can't keep up with Spencer and Leo for long, so I beg off for a break, and head down to meet Dave Richards, Alta's head of avalanche control, in the upstairs bar at the Goldminer's Daughter, the lodge at the bottom of the lift. The dark-paneled walls are plastered with black-and-white pictures of past dirtbags: the film stars who used to wash dishes here, and the ski pioneers who first skied these cliff bands. From the bar, you can watch people lap Collins through the roof-high plate glass windows, every turn on display.

Richards, who everyone calls Grom because he started patrolling at Alta and filming for places like TGR when he was a teenager, is here on his day off, casual in sneakers and jeans, although his manner is always intense. He drums his tattooed stump of a ring finger on the table while he talks, gruff voiced and full of eye contact. He says he's never wanted to be anywhere else, but that sometimes it's felt impossible to live here.

Grom was brought up on the tough-guy culture of ski towns. His dad was a patroller here, and he's both hypersmart and physically talented, which is why he eased into life as a professional skier so young. But he's also suffered serious trauma and PTSD from the death and injuries he's seen on the job, which have amplified his existing mental health battles.

After his best friend, a fellow patroller named William "Adam" Naisbitt, died by suicide in 2017, Grom decided to start talking about his own mental health issues, to try to normalize the idea that life in the mountains isn't ideal all the time. He'd come close to taking his own life, too. He's bipolar, which he says has helped his ski career, because being manic often helps him focus, but it's also made the dark parts darker. He thought he would lose his job when he started talking about it, because he assumed no one would trust their life, much less their avalanche bombs, to a guy who self-identified as "crazy." But he also thought sharing that he was struggling, and talking about the battle inside his brain, might break a little bit of the long-standing stigma, and that felt more important. It felt like the difference between life and death. In his line of work you're supposed to be able to handle anything, and if you can't, you're supposed to be able to drink it off, or ski it off, unfazed. That tough-guy posture is toxic, but it's germane to life in the mountains, where a certain amount of mental tenacity *is* crucial. "I call it the land of broken toys," Grom says. "This is where we all found each other and learned to play together. Some people who probably wouldn't make it work in the real world have found a home. It's a gathering of lost boys."

A lot of those boys wanted to talk about their brokenness, they just didn't have a good outlet, and when Grom started talking about it, he cracked something open. He's become a part-time therapist, particularly to young men who either can't afford to go to therapy (seasonal work doesn't usually come with great health insurance) or who are unwilling to talk to someone else about the mental load they're carrying, because they think it will make them look weak. "The reality isn't as glamorous as the myth," he says.

Going to the mountains can feel highly purposeful, especially when you have a mission, like a peak to climb or a mountain to patrol. But mountain towns also create a culture of scarcity. "Suicide rates have skyrocketed through the roof in ski towns, but there's a stigma, so people aren't talking about it," Grom says, throwing his arms wide to the half-full bar. "Look around here, everyone comes in and throws down their hat, and says, 'Best day ever!' and the person who feels like shit still has to say, 'Best day ever!' I think that's hard."

I know, full on, what that feels like. I have blinked back tears in the bar before, after a day of feeling inadequate, when everyone else seems elated. I have tried to disappear in the corner of the patrol room, because the alternative was feigning interest in accidents, which actually made me scared. I have spent plenty of days trying to control my breath on a too-steep boot pack, knees quaking. I have wondered what might be wrong with me, and why I continue to subject myself to my own abuses when something that is supposed to be pure pleasure starts to ache.

But I also know that the range of feelings is wide and memory is short. Today, for instance, is a very good day. The snow keeps falling, soft and smooth, and as the light gets low, it accentuates the shadows on the back of the bumps, blue-black in the troughs. After I leave Grom, I lap Collins till my quads shake.

SO BRAVE YOUNG AND HANDSOME

"I don't know who I am when it's not winter," my friend Sally Francklyn once told me, in the doldrums of a warm, slow November. Later that same season, Sal took a bad turn in the Jackson Hole backcountry. She fell 2,000 feet, crushed her helmet, cracked her skull, shattered her ankle, broke her back, and busted her brain, sustaining injuries that would shape the rest of her life. The first time I saw her after the accident, almost a month later, after they'd reattached the piece of her skull that let her swelling brain expand, she was thrashing and moaning, unable to control her body after the traumatic brain injury (TBI) she sustained in the fall. Nine years later, she can walk again and she's getting more independent, year by year, but the TBI scrambled her balance and her hearing, and skewered her sense of self. When we talk, it's halting and hard to go deep, and the most painful part is that she remembers how much she's lost.

Some people don't get to live out all the chances they might have taken. You go to the mountains and in the mountains people die, Grom says, and it's true in innumerable ways. That same winter thirty-four people, including several well-known guides and athletes, died in avalanches in the US alone. The month before Sally's fall, our former boss Megan Michelson—we all worked at *Skiing* magazine together—was a part of the Tunnel Creek avalanche, which killed three experienced skiers at Stevens Pass in Washington State. Megan says she still wakes up from dreams where she's digging bodies out of debris. Yes, you go to the mountains and in the mountains people die, but there is a difference between living with that idea in the abstract, and the bone-deep reality of loss. I'd lost friends and acquaintances in the mountains before, but that season shook me loose. That winter the danger became real.

The day that Sally fell I was patrolling at Arapahoe Basin when the phone rang. A friend from Jackson called me there. She was trying to track down Sal's parents. She gave me sketchy details of the accident and my head started to swim. All afternoon I imagined a series of increasingly bad scenarios. After the lifts closed I went to sweep the mountain with my friend Ryan, still not knowing exactly what had happened to her, or what was going to happen. As we traversed the boundary line, Ryan told me not to catastrophize, at least not yet. He said I couldn't make up stories or cast blame before I knew the details. We fall all the time, he said. And that's the thing, it's true. I fall all the time. I catch an edge and tumble, or misjudge a bump and faceplant. But when Sally fell I became obsessed with the randomness of risk. After that winter there were a series of years when I wasn't sure I still loved skiing. My lungs would shrink up every time a chute got narrow. I'd sidestep and back off of the kinds of steeps I used to dream about. My knees shook so hard. Skiing well takes aggressiveness, you need to be decisive and forward on your tips. Fear made me worse and it made me dangerous.

Before Sally fell, I was more afraid of embarrassing myself in the mountains than I was of dying. I had thought that I was invincible, my body elastic, until Sal shattered her skull and lay motionless in the Idaho Falls ICU for weeks. Now, Sally can Facetime. When she calls, her body and her voice are still jerky, and her speech is slow, although it's gotten better over the years. When her face comes up on my phone screen, I notice her super long lashes and the hole in her throat from the tracheotomy tube. Her skull was cut open, her brain was swelling, and we weren't sure she would come back.

Skiing is still a part of Sally's life. Sort of. The community at Copper Mountain, where she ski patrolled, still supports her as best they can, and sometimes she thinks about going back to Jackson, even though it's not realistic right now. It's still a piece of her character. I'm never quite clear how close you have to live to death to feel alive.

Snow is a changing, fragile substance, which accumulates in layers: a deep puffy storm, followed by an inch of rain. Wind crust followed by cold light flakes. Avalanches are a combination of three factors: a sliding surface, a slope steep enough to slide, and a trigger. Here in Utah—and in other high, dry parts of the Rockies—more often than not, there's a deep unbonded layer in that snowpack that could always slide, given a trigger.

It seems to happen the same way almost every season. The first thin snowfall covers the mountains in a crystalized layer of sugar and anticipation. Then it stops, like climatic clockwork, for a few weeks. That layer of unbonded snow is exposed to the air, which sucks out moisture, creating slippery, faceted snow crystals called depth hoar. It forms a perfect sliding surface. When the snow starts in earnest, that surface, which avalanche forecasters call a persistent weak layer, is at the very bottom, slick and unbonded, ready to slide. That's one of the constant hazards of skiing, you always know it's down there. Just how big it could break is a question of what comes in on top of it.

You need other factors to create an avalanche. The steep needs to be steep enough. Slopes less than around twenty-five degrees won't slide—that's why you only see ski patrollers controlling black diamond slopes—and ones steeper than sixty degrees won't hold much snow. And then you need a trigger, something as simple as an unwitting skier whose weight makes the slope crack.

Over the past ten years, an average of twenty-seven people have died from avalanches in the US each season. That's not many, all things considered, but in a relatively small community it can echo. Colorado and Utah account for more than half of those fatalities, and often in Utah those deaths occur in or near Little Cottonwood Canyon. "They died doing what they loved," is my least favorite mountain adage, but it's a frustratingly common refrain every time someone is buried in an avalanche or doesn't come back from a big expedition. When my friend Andre died in an avalanche on Wolf Creek Pass in 2012, on the back end of a season full of fatal slides and near deadly falls, I thought really hard about whether the lifestyle was worth it. Andre had, unquestionably, died doing what he loved. He'd been chasing winter season-to-season for years. He'd just bought a new snowmobile, which he rode into the sugar-crusted San Juans, where the snowpack released. But after that I got nervous when the snow moved under my feet.

Mountaineer Conrad Anker, a good friend of Bozeman counselor Timothy Tate, once wrote that, "We choose to play by the rules of the mountains because they are our calling. We accept the loss that strikes unaware in return for the bonds of friendship created by experiencing life in the majesty of nature. The intensity of the high alpine, guarded by wind and snow and ruled by gravity, is where we find these moments that define us as people. It is never easy to lose loved ones, particularly those with so much life to live." When it feels like a calling, the very thing that defines you, it's hard to back away.

I think about that idea of a calling any time I'm in avalanche terrain. Sometimes skiing is the only thing that can quiet my brain. I understand why Sally dropped into the couloir, or why it's so simple to convince yourself to ski into avalanche terrain.

By now I know how easily I could have made a fatal choice at so many different points.

I'm not sure if I'm allowed to call it dying young anymore, because I'm not really that young, and sometimes I think I've gotten too good at memorial services, but recently, a friend of a friend was dying slowly of brain cancer, and I realized I wasn't familiar with that gradual kind of death, which comes with at least a sliver of time to say goodbye. Before that, among my peers, I'd only been privy to surprise and immediacy: phone calls about a friend's fiancé, and a telescoping sense of dread that traveled through a bar when a well-known pro skier didn't come back from an expedition. Gut punch and gone. From what I know, death never feels fair, but there's a particular pain in knowing the person who died had some kind of choice in the manner. It's not that they loved it, it's that they put themselves in that situation on purpose, and the threat caught up with them.

Some of the most beautiful skiers I know never got to grow up. The first boy I met in college, John Nicoletta, was strong and even-eyed, so handsome that he could have been much more cocky than he was. In the early 2000s when backcountry skiing was on the rise, he was a telemark skier, which made me covet cable bindings and skin tracks. When I look back I can never quite untangle my risky decisions from my desires. Was I emulating the cool boys because I wanted to be with them or wanted to be one of them or both? I'm still not sure which romance is deeper.

Nicoletta moved to Aspen the year before I moved to Avon. He became friends with Pat and Tats, and slipped into that local crew. He was the one to secure rent at the Skier's Chalet, cementing himself in dirtbag royalty, and he started competing on the big mountain freeskiing circuit. In old videos you can see him compress like a spring when he hits big cliffs, before throwing himself powerfully into the void.

He was starting to make a name for himself as a skier, doing well in those big mountain competitions, and coaching younger kids. He even made the cover of *SKI* magazine. The photo is of him holding one perfect powder turn. In my mind, Nicoletta is still frozen in that golden light. In 2008, he fell during his competition run at the US World Freeskiing Championships in Alaska, and kept falling. He was dead by the time his body stopped moving. He was twenty-seven.

In Aspen, Pat took me to the shrine they built for him after he died. It was springlike, warm with a fresh coating of soft snow when we tracked into the woods, and dusted off the bench with his name carved into the seat. I leafed through laminated notes and photos tacked to the aspens, while Pat shoveled off the roof of a shack, which is propped up on old skis. It's been more than a decade since he died, and everything at the shrine looked a little faded, even when we brushed off the snow, but the loss still aches. It's Nicoletta, and all the ways he didn't get to grow up, but it's also the weird glamour of freezing the person in time and glorifying the risks they took in life, even if it took them over the edge. In Jackson Hole, Doug Coombs is still called the best skier in the Air Force, even though he died in 2006. In Squaw Valley, Shane McConkey, the star of the early Matchstick Productions movies, is still worshipped, but his daughter is growing up without a dad.

The spring after Sally's accident, when I was still reeling, another mutual friend, Eric Henderson, took me into Once is Enough, so I could see the steep-walled couloir where she'd fallen. I didn't want to go, and the creep of my obstinacy surprised me. I sucked back tears when he told me I should know the place, because it would help me come to grips with what had happened. I refused to drop in from the top, but Hende gently forced me to ski out to the bottom of the chute, so I could have context. It hadn't snowed recently, it was warm and muggy in Jackson, and the snow was slushy under our skis as we skinned out under the Hobacks, and around the curve of Cody Peak. The chute cuts down the mountain's slanted shoulder. Hende took a picture of me that day, standing in front of the narrow hallway of snow, trying to gauge the steepness, trying to understand for myself if I would have dropped that day. My goggles were fogged and full of hot tears as I tried to untangle the trauma of loss.

If you get deep into skiing, eventually you have to acknowledge that the thing you love can kill the people you love. To cope we turn deaths into hero stories or we go silent. You rarely see the accidents, back-offs, and close calls in ski movies, even though athletes say it happens all the time. Mountain culture prizes shaking off tragedy. But if you're actively ignoring the brutal parts, especially when they keep happening, the trauma will catch you.

Grom knows exactly what it feels like. He still flashes back to the nonstop storm season of 2005. He says that as an avalanche professional, it was the kind of year you dream about, but hope to never see again. In the depths of a class 5 avalanche cycle he dug up four dead bodies in the course of three weeks, and he says the image of the first one remains impossible to shake. "He was wearing a green jacket, and the first body part we got to was his left hand with a gold wedding band on it," Grom says.

Instead of talking about it, Grom coped by drinking and making jokes, and throwing himself into dangerous situations again. He ignored the panic and tried to tell himself that he was fine. "Because of that, it's very, very difficult to stand up and say, 'I'm struggling here when I'm supposed to be tough,'" Grom says. "A 'buck up, Chuck' attitude is pervasive in mountain communities and mountain cultures."

Trauma can look like a lot of different things in the mountains, from watching a ski partner get seriously injured, to helping with a body recovery. I often feel like I shouldn't articulate my fears, because I'd betray weakness. Such an admission would mean that I couldn't handle the stress.

Grom is trying to break that association between fear, stress, and weakness, and he has science behind him. The National Institute of Health defines a traumatic event as a shocking, scary, or dangerous experience that affects someone emotionally. Trauma impacts everyone differently. Grom says some people won't be fazed by a rescue or a death, but that others will suffer long-term PTSD, and it can be almost impossible to say when that will happen, whether it's on the first avalanche burial, or the fourth or never. For me, Sally's fall is woven into my neural patterns. I can't not think about her when I ski.

The impacts of trauma can manifest in many ways, from exhaustion to hypervigilance, and it can be hard to tell where one begins and another ends, or where something like stoniness goes from being helpful to being toxic. Grom, who has gone through periods where his anxiety is so bad that he can't sit with his back to the door, says vigilance is crucial to being safe in avalanche terrain, but his uncontrollable hypervigilance can be brutal.

According to the NIH, the best way to address trauma in real time is to talk about it with people you trust, which is why Grom has been so insistent about telling his story and asking his coworkers if they're OK, even if they brush him off. The number one predictor of how someone will recover after a traumatic event is how safe they feel in their relationships. After a while coping mechanisms, like that beer with friends, or blowing off steam with exercise, stop working. Or worse, they exacerbate the problem. But talking through guilt, shame, and grief, especially in a cognitive behavioral therapy setting, can help. Knowing why you feel bad peels back the pressure of feeling bad itself. Talking about it takes away some of the shame. "Once people understand that's what is happening to them they literally can see the way out," Grom says.

EXTREME EXTREMES

This winter on the road, I spent a portion of nearly every night I was in the mountains at the bar and a good portion of most days with alcohol in my system. On one day at Alta, Spencer took me to his liquor box. It's a hidden birdhouse-like crate, nailed to the back of a tree deep in the woods, that he keeps locked and stocked with Fireball, Bulleit, and pornographic cartoons. He poured me a shot of Rumple Minze before we headed downhill. Then I had a beer with Grom in the GMD, and another on the lift. After that, we went to the P-Dog, the classic bar at the Peruvian Lodge, and posted up at the old wood counter, pushing back PBRs.

There's a mountain town trope that DUIs are common. When I lived in the Vail Valley the joke was that you weren't a local until you got one, but the jest belies the destruction of hard partying, which is prevalent here. According to the *American Journal of Public Health*, ski town counties have some of the highest rates of adult drinking in the country. My doctor told me recently that I drink too much, and I balked because that didn't seem possible. "I know so many people who drink so much more than me," I blurted at her, my logic already twisting as the words tumbled out. When I lived in the mountains I used to start drinking at 5:00 p.m., or whenever we got to the locker room, and keep rolling from there. At first, I wanted to prove that I could keep up, even if my skiing couldn't. I was much more lucky than good or right, as I have been in so many situations. But when you're looking at your friends and they all seem like they can drink lots without faltering, it's easy to think eight beers in a night is normal. As with avalanches, not dying doesn't mean you were smart. So often it means you got lucky.

When I returned to the mountains it was hard not to slide right back into the pattern of drinking excessively. We drank in the patrol locker room and on the ridge top after a hike. Someone cracks one on the chair, and another in the woods. And that's before you even go to the bar, or the parking lot, and actually start drinking. You go to après every night, because that's your third place.

Impulsivity, one of the tenets of sensation seeking, is also tied to drug and alcohol use, and Cynthia Thomson, the Canadian health researcher who studied risk in Whistler, found that the people in her studies who scored higher than average for impulsivity both skied and partied hard. The drive to push limits is internal, but it's reinforced by culture, and once you've elevated your baseline it's hard to come back. Bozeman-based psychotherapist Timothy Tate says people frequently come into his practice suffering from some form of alcohol or drug addiction or heavy use. Often it's because they have been in a tragic accident and are feeling lost and confused and angry. They don't have another coping mechanism for dealing with trauma, and they didn't even know they might need one.

Drinking socially can be therapeutic, and build camaraderie, but, like skiing, it can also become a dangerous way to avoid real life. "There's a lot of good things about mountain communities, but there's not a lot going on at night besides the bar," Grom tells me, when we're drinking at the GMD. "If you're struggling with mental illness, social stimulation is excellent, but if you mix that with alcohol every night, you're just driving yourself into the hole."

Spencer is a proud proprietor of the liquor box. He keeps it filled, but he doesn't drink anymore. When I first met him, his eyes were often fuzzy behind his sunglasses, and at rowdy social gatherings he was usually driving the debauchery. After odd jobbing and ski coaching post-college, he got a job running marketing for Pit Viper, a sunglasses company that brands itself as part of the party. It was his responsibility to propagate that image and he was very good at his job. He'd be the last person at the bar, the one rollerblading into the hot tub at the end of the night, or starting the lift line karaoke. I've seen him naked in public more than almost anyone I know.

He said that he liked being that guy and he got attention for it. His boss thought it was hilarious when he got embarrassingly drunk at work events. He saw the people he looked up to, like the pro skiers he'd admired since he was a kid, raging until 3:00 a.m., and he wanted to follow their lead. It seemed normal to his peers, who went to extremes in a lot of different ways. "My philosophy is that we're in this one percent of thrill seekers," he says. "We chase these peaks of fun and excitement, but it also means we're all kind of crazy people."

It's hard to be at eleven all the time. And for Spencer, all the partying started to feel shameful. He said he would try to limit himself to just drinking on weekends, or would try to take weeks off, but it was hard when his job and his social circle commenced at the bar, and when he was expected to bring the ruckus. The scene enforced his tendencies toward excess and thrill. "The year before I stopped, I noticed myself starting to fuck up more often, it was taking a bigger toll on me. I had all this shame and I was out of control," he says.

And then he ended up in jail, after he passed out on a stranger's lawn and was charged with a public intoxication misdemeanor. The next day, when he started talking through what happened with his dad, he realized exactly how skewed his life had become. "He asked me, 'How often do you get blackout drunk?' and in telling my dad that, I realized I should never have gotten that bad," Spencer says.

He decided to stop drinking for a year, and see if he could still be in the ski world without it, or if his work and life and friendships would change. And some of them did, but he says he's unequivocally happier now when he's not sucked into substance use. His relationships are stronger, his skiing is better. "I don't think that any of my friends disrespect my choice, but I have stopped seeing a lot of people I used to see all the time," he says.

Spencer says he has a certain ability to hang still. He's now okay being the sober guy at the party. He's past the year mark and still going. "I think, in a completely not humble way, that's a unique skill or power, to be able to do that," he says. And I think it is. I don't think I'd be able to do it. I've seen him stay up, late night, at events when everyone else is smashed, still having fun. Even here, in the tree-side bar he's created, he's still enabling other people's good times.

Ryan Burke, the therapist in Jackson Hole, says that sobriety is particularly difficult in mountain towns. He also struggled with substance use while working as an addiction counselor. He knows how hard it can be to break out of that free fall, because the culture supports the idea that you can go from first chair to last call, and then get up and do it again. As an addiction counselor in a mountain town, he sees lots of overlap between adrenaline sports and drugs and alcohol, both in terms of increased tolerance, and the way people continue to use despite negative consequences. Many ski bums he encounters don't have the personalities to say no. "I don't think people in other places understand that here it's very normal to say 'I'm gonna ski pow-der all day, have some alcohol, maybe do some coke, then get up and do it again,'" he says. "And when we're going hard, we often don't see the negative consequences before they're bad."

In many mountain towns a dangerous self-medicating drug habit rumbles just beneath the surface, too. Burke says that he has clients who use drugs to overcome injury. They get prescribed painkillers after an accident or surgery, then keep using, because it helps sustain the lifestyle. "People get all righteous about it, but if they were injured, I think they'd do it, too," Burke says. "You lose your tribe, you lose your ability to cope, you lose ev-erything that's important to you. You take away someone's main coping skill, it gets ugly." Burke says self-medicating is basically normalized in mountain towns, and the repercussions go un-spoken. "People who die in avalanches get put on a pedestal, people who die of addiction get ignored," Burke says. "We ig-nore what we don't like to hear."

GOING DARK

What happens when you get to a point where you can't ignore the dark parts? In March of 2017, the Alta community was blindsided by two deaths by suicide in four days, Donald Brantley, who worked at the Rustler Lodge, and Adam Naisbitt, Grom's best friend who had been an avalanche forecaster and ski patroller.

Grom says that last time, when he cracked, it was because of Adam. "When Adam hung himself it was just this huge blow to this community," he says. "People knew he was sad, but people didn't have a clue, and a lot of people are still spun by it, because they don't understand why, or what to do about it." Grom ended up in the hospital, after he, too, attempted to take his own life. He realized he had to change some things fast.

Sometimes the mental risks are worse than the physical. All those factors: the isolation, the constant need to prove yourself, the trauma of people dying, the stress of being poor and having unstable housing, the unhealthy burn of self-medication, and the incessant grind of a brain that won't calm down. It all piles up. Sometimes living the dream doesn't feel like it's worth living anymore.

Ernest Hemingway died by suicide in Sun Valley, and Hunter S. Thompson killed himself in Aspen, in 2005, one of five similar deaths in the valley in a two-month period. Celebrities shouldn't be trend forecasters, but in those cases they put a spotlight on something happening in mountain towns. In what is purported to be paradise, people are killing themselves at alarmingly high rates, and it's getting worse.

According to the Centers for Disease Control and Prevention, someone who lives in Wyoming is five times more likely to take his own life than someone who lives in Washington, DC. Salt Lake County, home to Alta, has twice the national average number of suicides. Researchers who study depression call the region that stretches across the Rockies the Suicide Belt because rates are so high. Those rates have gone up more than thirty percent in half of US states since 1999, and rates in mountain states like Montana, Idaho, Utah, and Wyoming are growing the fastest.

In 2017, the Utah Medical Examiner's Office hired sociologist Michael Staley as the state's first suicide researcher to try to understand what was happening. Suicide rates were essentially stable for three decades, but they've been increasing since 2014. It was a public health crisis.

According to the Center for Disease Control, there are 300 factors for suicide. They include isolation and comparison along with lack of relationships and lack of access to mental health care. They also include easy access to guns. Being male and stubborn. Getting older. Those factors track across the wide range of the Mountain West, and they're highly concentrated in ski towns.

Staley says that he doesn't have exact data on ski towns, especially because numbers on attempts versus completion aren't very good, but that demographic trends in suicides overlap with mountain living. Chief among them is that men are driving rising suicide rates. Nationally men account for seventy percent of suicides—in Utah it's seventy-six percent—because men are less likely to seek help. Additionally, he's found that about a third of people have alcohol in their system, and that mobility and lack of community ties are both factors, as is qualitative disenfranchisement among white men. "I don't have any statistic to say this many people have died by suicide in ski towns, but if you look at the life of a ski bum, the reality is that they're working hard to make ends meet and living in the world when you're serving the wealthy, which you can't afford, there's a big disparity between you and them," he says.

I've felt the ache of that disparity in so many different ways: frustration at my inability to live at the same standard as my friends, even though I didn't realize they had the boost of secret family money, or some other sneaky cushion. Self-hatred when I couldn't keep up or hang. I spent a lot of time questioning why I couldn't just be happy and relaxed and living in the moment like everyone else seemed to be. The feeling of incompetence is a big part of what drove me away from life as a skid. That perceived failure can trigger depression for people who are prone to it. For a lot of people—not exclusive to ski towns—shame stops them from getting any kind of help. Plus, the divide is worse when you feel bad and everyone else seems to feel good. "There's a term called relative deprivation, which is why you feel even worse if you feel bad in a ski town. If everyone is in a shack, you're happy, but if you're in a shack and everyone else is in a mansion, that hurts," Ryan Burke adds.

On top of that, a growing body of research is examining the mountains themselves. A 2019 study from Auburn University, which looked at county-level suicide, found that for every increase of 100 meters in altitude, suicide rates increase by 0.4 per 100,000. In 2008, Dr. Perry Renshaw, a biophysicist at the University of Utah who had recently moved into the state and began to feel down, started looking at why. He developed a theory that suicide rates are correlated to the lack of oxygen at altitude, because hypoxia has been linked to decreased serotonin, the neurotransmitter most directly tied to well-being and happiness. If you take an antidepressant, including commonly prescribed ones like Lexapro, Zoloft, and Prozac, there's a good chance the drug is what's called a Selective Serotonin Reuptake Inhibitor, which keeps serotonin levels high in your synapses by preventing it from being absorbed. Having more serotonin in your body generally keeps you happier and more stable.

Renshaw has found that in animals, serotonin levels can decline by as much as thirty percent in a day at altitude. He says that similar drop in a human brain could lead to devastating depression, and low levels of serotonin byproducts have been linked to increased suicide risk. Just like dopamine, everyone processes serotonin differently. That's why one person might be affected by the altitude, and another might not be fazed.

Spencer says sometimes he looks at pictures of previous seasons and they feel fake because they don't capture the tension. And it gets worse when those images are filtered through social media. Timothy Tate recognizes a pervading millennial idea that we're all supposed to be shiny, individual snowflakes who can do anything we put our minds to. Social media stresses the comparison. Tate says that to work through it you have to conquer the constant fear of being left behind.

When I talk about the years I spent working menial, seasonal jobs, and living in thin-walled apartments, just to be close to the mountains, I tend to gloss over the hard parts. I give in to the story I think I should tell. "It was the best," I say. "All we did was ski and party and go on adventures. I didn't have anything to worry about." And really I didn't have much to worry about. I was making rent. I had the things I thought I wanted: a modicum of insider knowledge, a web of friends, and a ski patroller boyfriend. Maybe the occasional tourist got mad at me for ruining their vacation, but the big-picture stresses were insignificant. In spite of what should have been easy I was anxious and down. I was scared of my own darkness. When I read back through my journals from that time, my writing voice is high-pitched, even on paper. Worried if I was doing it right, if I was doing anything right. Was I in the right relationship?

The right place? I had spiraling guilt about not contributing to society, and I was missionless, in exactly the way that Tate says is emotionally devastating for many ski bums. Maybe if I was better it would be better, I can see myself thinking when I read between the lines of those old notebooks.

Things that might be red flags in other places—like daily drinking or obsessive exercising—seem fine in the mountains, where everyone is a little haywire and going to extremes. Your social circle might not think to catch you when you slip into free fall. Maybe you lose some friends in an avalanche and wonder why it wasn't you who got swept away. Maybe you keep dipping a little deeper into debt, because you can't quite make enough to stay ahead of your rising rent. Your body breaks down as you get older, which means you can't hit those dopamine highs in the same way. Your job gets harder when your joints get stiffer. Maybe you get hurt, which means you lose your physical outlet and your daily lift line peer group, so you start going to the bar more, to see people, and to get a sense of release. Maybe it starts to get bad, and your friends let it slide when you pass out places you shouldn't, because even though you see them every day, they aren't the kind of long-term friends who can call you out on shit. Maybe you've got painkillers, too, and they dull the drive. Maybe you can't stop feeling bad because your life isn't shaping up to be what you thought it would.

Timothy Tate says those thoughts often come from an existential crisis about meaning and identity. "It's different for everyone, but you have to face your shadow. It's nothing you can tread lightly or behavior modify your way through," he says. "It's a big question: How are you going to reconcile your life with your choices and your static identity when your sense of meaning and value and where you fit in is so tied to what you can do? That's such a breakable thing."

★ ★ ★

One night at Alta I end up at the P-Dog, the bar at the Peruvian Lodge. I show up early, while the lifts are still turning and the dusty room is full of sun streaks. I grab a beer with Lee Cohen, a wide-grinned ski bum who has been shooting ski photos since he moved here from Colorado in the '70s, after a stint spent sleeping in his car near A-Basin. His tightly framed powder shots have become iconic. If you have a mental picture of the bottomless fluff of the Wasatch Range, it might be based on one of his snow-caked photos. Upstairs in the P-Dog, we munch bar mix as the light drops lower, and he tells me that even though it's hard to make it work as an old-school photographer in a digital world, even as the canyon changes around him, and his peers peel off for other, less Peter Pan-y towns, he's never been able to leave. There's something specific about the light here that he's never found anywhere else. And now he's raised a kid here, a second-generation skid, who is trying to carve out his own path in skiing.

The lifts stop running, the tourists stream in, and then the locals do, too, once they're done with work on the mountain. Grom comes in to have a beer with two fellow patrollers. A mediocre band starts up in the corner.

Sara Gibbs, Grom's ex-wife, is the bartender at the P-Dog, and she's cracking beers with a watchful eye as the different social circles cycle around each other. People are mostly still in their ski gear, a lot of dudes with long hair, one woman wobbly on mushrooms. Sara says she can tell how the kids are doing pretty quickly when they come into the bar. She calls them all kids, even the people who have been here for decades, and it's become her unofficial duty to keep asking how they're doing.

In addition to working here at the Peruvian, Sara is the executive director of Alta Community Enrichment, which, among other things, is the de facto mental health center in Little Cottonwood. She didn't sign up to do it initially, but after Adam

Naisbitt and Donald Brantley died by suicide, she helped ACE
start a program to address suicide threats and mental health in
the canyon, because things were getting scary. "I can see when
people are struggling," she says. "Adam was a close friend of
mine and I was with Grom for twelve years. Those experiences
with Grom were super raw—it was probably the demise of our
marriage—but I knew something had to happen, so we said,
let's start this conversation about mental health." She thinks that
conversation is key to being able to thrive in a place like this.
And she's made it her mission to give people space and support.

She didn't have a therapy or counseling background, but she
started to see common behaviors, and notice the need. "I got
trained in how to help with suicide threats, I kind of became
a go-to by default," she says. In addition to getting called in
when there are attempts or suicidal ideation, she's also trying to
advocate for pre-traumatic stress education, and to get people
talking about mental health before things get bad. Like Grom,
she's trying to get the good and hard parts out in the open. She
wants to show the less-shiny side, and explain why people stay,
despite it. She knows why people come to the mountains, and
why they find it surprisingly rough as they settle in.

"I use Grom as an example of people who are struggling with
mental health issues, but who are also really strong athletes,"
she says. "It's really common. We get these kids up here, they
realize how good it feels, and they want to go higher and faster
and more and more, getting endorphins, filling that void." Sara
doesn't necessarily want to be behind the bar anymore. It's si-
multaneously communal and isolating. But her role gives her
purpose. So for now she plans to stay.

INTO THE MYSTIC

Spencer is part of a crew of younger guys who are constants in Little Cottonwood. It's a loose conglomerate of skiers that changes slightly from season to season, but if you show up at Alta, you'll usually have someone to ski with. One of those people is photographer John Howland, who is tall and lanky with a low New England accent. He has a disheveled look that belies his smooth skiing. As a photographer, he can capture action shots. A tail of billowing snow. The compression of a turn. He's the next coming of Lee Cohen. John moved out west after college with a group of friends from Vermont, and stopped in Big Sky for a season before ending up here, in the Cottonwoods. He's got buddies in all the canyons, and an uncanny ability to be wherever is getting the most snow. But he still says he gets hit by waves of unease about where his path is headed.

From the outside, he knows that his life looks ideal. He lives at the base of the canyon, he gets to ski all the time, and over the course of a couple of years he went from tuning skis to shooting photos for places like *SKI* magazine. He says that as a kid he would have felt ridiculous imagining his current life, which is one of the reasons why he can't give it up. But he says he struggles on days when he doesn't want to go out and ski.

He's worried about alienating himself over a temporary lack of stoke. He's pretty sure other people feel like he does—when he's brought up his anxiety and depression, it's resonated—but he says it's terrifying to try to get off the ride sometimes. "No one ever acknowledges the difficulty," he says. "I think the Salt Lake City area in general is hard, it's the most athletically motivated and competitive place I've ever been. Everyone solves their problems by getting rad, but it's not the answer. And when you're doing something that you love, but it doesn't feel good, that's really frustrating." He feels stuck in a flattened identity. He's the next generation of ski bums, and right now he's dealing with what that means.

His battle for clarity is essentially evolutionary psychology, according to Timothy Tate. Tate says that identity, pleasure, and risk, and the ways they show up together, come from how we've evolved and how we try to fit into society—whether in Alta or Atlanta. Someone like Howland, who has been obsessed with skiing since he was a kid, would lose a big part of his identity if he dialed back on skiing.

When he thinks about getting older in a ski town, Ryan Burke says he thinks a lot about psychologist Erik Erikson's psychosocial stages of development, a theory about how your personality develops over your lifetime. "In your late thirties and forties you get to a point where you're battling stagnation against generativity," he says. "That's what I see a lot of times here. The day-to-day pleasure is still there, but you get to forty and you're like, 'What am I contributing?' You need to feel like your life is bringing you somewhere, depression comes from a lack of that."

I know that I seem more like myself in the mountains, that there's a piece missing when I am gone. The need to be there feels more like a compulsion than a choice. I often feel the same about writing. I could be doing a lot of other, more purposeful, more lucrative things, I'm sure, but somehow life has slung me into this rut, and I can't and won't get out. I need to feel connected to the past. I need a sense of purpose.

"That question of living the dream, and what it means in America right now, it's super relevant," Auden Schendler told me in Aspen. "There has to be a reason to stay."

Just down the canyon, at Snowbird, which splits a ridge with Alta and spills over the other side of Mt. Baldy, a crew of older guys stream into the Forklift restaurant. They're snow-caked and sweaty, ditching their helmets and jackets at the door. It's 10:45 a.m. Same time every day, every winter, for forty-three years. Scott, Pete, Neil the old ski area archivist, Guru Dave, and seven others are sitting around in the sparkly late-morning light, drinking coffee refills out of mugs with their faces on them—a pastiche of their old pass pictures, which document nearly half a century of winters. In the pictures, their hairlines fade back, and turtlenecks and mustaches track the trends of the '70s and '80s, and the ways they've settled into their lives in the mountains, grayer now, mustached still. They tell me they used to date the waitresses, and take them skiing. "Now they carbon date us," Guru Dave says, cackling, as the table orders bagels with cream cheese, and eggs and toast. Some of them have wives, but all of them have the kinds of jobs where they can take off on winter mornings. This congregation rules over other obligations.

Maybe because I'm there, they tell old stories about disco days in the Tram Club, the bar in the basement, and about how they ended up here, in Little Cottonwood. They tell me about all the phases of cool—the bump skiers and the freeskiers and the big mountain bros—and how they've stuck around through all of that. Even though the canyon has become crowded, they're still there, chasing powder, but trying to go easy on their knees.

Some of them go home after breakfast but the Guru and I head back out into the biting, glittery cold. The snow squeaks under our skis, and I follow him on his hunt for soft turns. His friends

gave him the moniker Guru Dave way back when, because he
was constantly expounding on weather conditions, and he had
an uncanny feel for what a storm might do. That unrelenting
obsessiveness has turned into a daily forecast (he skis about 130
days a year), a modicum of local celebrityhood, a book about
Snowbird, and a life philosophy based on the pull of these particular mountains.

After forty years of tracking snowfall he has a mental map of
all the wind-loaded tree pockets, and the cornices shaded from
the sun. So we zigzag the mountain, traversing to find slices of
soft snow. As the sun slides higher we drop over the backside
into Mineral Basin, which holds a lucid, purple light, the wind-
blown snow crystals refracting.

The light flickers in and out as I chase Guru Dave down the
rollovers, trying to mimic his rounded turns. It's cold enough
that I start to get waxy white spots on my cheeks and nose, but
the snow is soft and spreadable, bump free, and firm enough to
push. Dave's been here since November 1976, when he bought a
one-way ticket to Salt Lake City. He says the path that brought
him here just kept narrowing until there was nowhere else he
could be. It sounds hokey when he says it—and I know it does
when I do, too—but on the chair he explains why he felt pulled
in, and his rationale resonates with me. His summertime day
job is paving roads, but he studied comparative religion, and he
says his need to be here is a part of his philosophy, and of trying to find a place in the world where environmental energy
lines connect. When we come off the chair in Mineral Basin the
light is fractal and glinting off the snow. "That's the energy," he
says, as I squint out over the ridgeline. The sheer mountains are
a massif of Tintic Quartzite, and he says that geology is what
creates the force that makes this place feel special.

He says if you let it, you can feel the lines of energy flow
through your body when you're rushing downhill, in a way that
only skiing allows you to do. He's searching for stillness in his

turns. He tries to think about keeping his hands motionless, he tells me, and to live in his body, connected to the mountain. Grace for him is taking that energy and trying to project it. He accelerates over the lip of a rollaway and I follow, trying to turn off my brain, trying to let the arc of motion be enough.

The theory of person-environment fit says that you might be happier and more fulfilled in certain places because of your temperament, values, and goals. That your characteristics might match up best with a landscape or a location. It's often applied to business structure, and organizational psychology, but could just as well apply to life in the mountains. As I follow Guru Dave, it's hard to picture him anywhere else.

Two things Tim Tate told me have stuck in my brain all winter. The first is that idea of soul. That there is something in your internal makeup that drives your need to be in the mountains. The second is something he tells a lot of the people who come to him suffering, unsure of their life choices, and frayed by stress and loss. "We need elders," he says. We need to stick around long enough to get old, and to show each other how to survive and succeed, even if our definition of success doesn't look like anyone else's. This is Auden's idea of purpose, and drive, the skinny edge of finding meaning in our own iconoclastic ways. It's Pammy carving out her space on the ridge, Mikey's gravel-scattered laughter in the trees. If you're the kind of person who is drawn to skiing, you've probably seen your own piece of that, too.

The spiked peaks of Snowbird ease up at the bottom of the valley. Still steep, but softer here. The trail we chose slopes and pitches into a natural half pipe. I follow the Guru down to the pocket, tracking his turns into the deepest snow, nowhere we could be right now but here.

CONCLUSION:
LIVING OR GIVING UP
THE DREAM

While I was in Utah, storms were piling up in the southern Rockies. Last winter was dry as dust down there, but this year has brought so much moisture that avalanche paths are sliding to historic depths, sending walls of snow crashing over roads, and ripping down stands of trees. At this point in the season I'm consumed by storm chasing and storm watching. So from Little Cottonwood I point the wagon to the heart of the steep, rarely-snowy-enough San Juan Mountains. I drive over Red Mountain Pass, arguably the most treacherous stretch of road in the country, to Silverton, Colorado, one of the last little ski towns.

My friend Jeff Davis, an avalanche forecaster I met through Sally, lives there. He forecasts for the highway, a highly coveted snow-science job, because it's a stable government gig that also lets him spend a lot of time on his skis. In the fall, when I first announced that I'd be on the road for the winter, he said I could come crash with him whenever. At this point I am sunk back into the informal economy of couch surfing, and not shy about using all my invites, so I text that I'm heading his way. Can I stay? Can we ski? And even though it's not particularly convenient, because the serial storms keep his radio going off all night, and there's someone already sleeping on that futon that he offered up, he says yes.

Jeff's house is an old mining shack, tight in the joints and rickety in the rafters, almost invisible in the roof-high snow-banks. There are three trucks pulled in out front, and two ski guides sleeping on the floor when I roll in with my open case of cheap beer. The fire is hot, and Abe the dog is curled up in the blast. The floor guys are Coop and Ryan, and they're in town for a week to take a level three avalanche class that Jeff is help-ing teach. The boys are better at taking care of themselves than I am. They brought vegetables and tea instead of a half rack of Tecate, and over the next few days they make us a series of big hearty meals: curries and massive breakfast scrambles, while Jeff keeps his eye on the forecast, and on the highway. I unroll my sleeping bag in a corner of the living room, stoking the fire and boiling the kettle.

Silverton is a specific kind of grimy paradise. It's a five-block-wide former mining town, cuted up for tourists, who mainly come in on the train in the summer, riding the rails from Du-rango. In the winter the churn of visitors slows down and the dirtbags take over. There are two roads in—the ones Jeff is re-sponsible for keeping open—and both of them are treacherous. You come here on purpose because the only thing going on is skiing. There's a dinky four-trail ski hill in town, super-steep Kendall Mountain, but the real draw is Silverton Mountain, just up the road above town. Silverton was the last real ski hill built in the US. It's a one-lift, mostly-guided mountain, which prides itself on rugged terrain, and not much else. The Silver-ton base lodge is a canvas tent that smells like sweated-through Gore-Tex. The bar is a keg in a bucket. The bathrooms are the port-o-potties, or, more often, the trees.

There is pretension and posturing, because it's a ski town, but it feels toned down and mellow here. Ryan, who patrols on the other side of the mountains at Telluride, says rich people only want remoteness if it's easy, which Silverton is not. It's the kind of scrappy small-scale community where you can stumble to the bar in your ski pants and always see someone you know, or find someone to talk to. Silverton isn't perfect, but there's an unrepeatable alchemy here in the isolation and simplicity.

A storm is supposed to roll in over the next few days, heavy and wet, so Jeff is monitoring the forecasts, watching the radar for the green swirls of moisture building over the mountains to the west. The highway radio by his desk is buzzing. The passes might close and I might get stuck here for a while, but for tonight there's not much to do besides settle in, take the bowl of soup that's handed to you, and talk about skiing.

The three of them are part of that rare group in which their personal, professional, and physical interests all intersect. Their lives are an almost-overlapped Venn. Their peers are their friends. Their deep nerdiness about weather makes them cool here. Their antsy need to move their bodies and their antisocial tendencies make them good at their jobs. Those bodies might break down at some point. Eventually snuggling with Abe on a boot-strewn floor might feel unbearably lonely, and those jobs, which come with a dose of seasonal poverty and instability, might lose their shine, but for now, they have a network that comes with open doors in desirable places, and a drilled down sense of purpose.

Another guide comes by with another case of beer, and we sit around joking about high-altitude fart tennis, and telling stories about past trips, and past storms, and the people we've met along the way. And I ask the question I've been trying to answer in every town: Can you still be a ski bum? Should you try?

And they say yes and no and it depends, and maybe for a while. Jeff, who is Midwest-chatty with a corona of blond curls and a face-wide smile, ended up here after stints in Telluride and Summit County, and he says life in Silverton is good, generally, unless you're interested in dating, or fresh vegetables, or if you get stressed out by the echo chamber of a very small town. It can get weird and lonely, he says, and sometimes he worries that he's drilled too deep into the physical parts of his life. He says he thinks about it like a stove, you can't have all the burners—recreation and romance and work and friends—going at once. He worries he might be burning too hot on the wrong flame. He knows that the lone wolves with their antisocial behavior sometimes settle into patterns that don't serve them. He sees fifty-year-old dudes alone here, and he doesn't want that. But for now the rhythm suits him. The San Juans stretch out for an unwalkable number of miles, he has a job that feels purposeful and interesting, he can take Abe the dog skiing right behind the house, and right now his rooms are full of friends.

The next day the storm is still hovering on the west side of the mountains, so I tag along on a ski tour with Jeff and another forecaster, Chris Bilbray. They know their work will be crazy for the next few days, so they're taking the time to stretch their legs before the region starts to get hammered by snow. We ski what they call the Soul Trees, a wedge of terrain right off of the base of Red Mountain Pass. The trail kicks up steeply almost immediately, grabbing the air out of my lungs, so I walk slowly to 11,300 feet, making slippery kick turns in the trees, letting the professionals set the track through the wind slap and the sun crust. I ask questions to try to keep them talking so I can try to keep up, but I don't need to do much because the two of them can talk for hours about snow: the exact crystal structure, moisture content, and depth hoar, the semantic difference between persistent weak layers and deep persistent weak layers.

We stop for snacks on the ridgeline, tucking into the lee of a rock, out of the wind. Jeff tries to point out peaks in the distance, but we can hardly see through the slop. The storm is rolling in, the skies are soupy gray in the distance, and the wind has picked up. Chris digs a snow pit, to look at the record of the winter crystalized in layers, and so he can say that today was a forecasting workday and then we switch our jackets, spin our bindings into ski mode, and start to make our way downhill. We drop off the shoulder of the ridge into an east-facing gully, linking turns between the soft snow on the north facing side, and the sun-crunched crust on the other, trying to find the soft parts, searching.

Maybe this is the vision quest Marcus saw: the hunt for some skinny moments of grace, tucked in with people who have made these mountains their homes. At this point in the winter I've bought in. I'm paying much more attention to the weather than I normally do, much less to the news. I am making my plans around storms. I have been ping-ponging around the Rockies for the better part of a season, since I came down from Jackson, heading back and forth to Utah, and up and down I-70, chasing storms, trying places on, living back in the spaces of my memories. I'm trying to lean in to what feels good, and figure out why it does, but that hunt has left me so swollen by nostalgia that I'm not sure if I can trust my judgment anymore.

Nostalgia has roots in Greek, it means painful longing for home, but right now the confusing part is that I'm not exactly sure what home is, or where. If it's a feeling or a group of people or a place.

Snow is too obvious a metaphor for instability. This year is good here, but next year might not be. At some point we might have to let go of the places we love, but, as the painful trope goes, is it better to have committed to that love, even for a little, than to never have known it at all? When I tick back through all the people I've spent time with this winter, from Pammy to Pat, they all have a purpose that keeps them in the mountains.

That night, in the rise of the incoming storm I walk down the road to the Avalanche Brewing Company to meet up with

Jen Brill and John Shocklee. Silverton is a ski town in big part because of them, and so I want to know about their vision, and how it holds up.

Silverton Mountain really was a dream. Jen's husband Aaron came here in the late '90s, looking for somewhere he could start a ski hill. He and Jen, who were in their twenties at the time, had been living in Montana, scouting mountains there. They liked the feel of small single-lift resorts, like the club fields they'd visited in New Zealand, and they wanted to make something similar, more about the skiing and less about the resort, dirt-baggy and simple. It took them four years to settle on Silverton, where the steep, snowy stretch of mountains held skiable terrain. After they found a place that felt like a good fit, they had to wade through a legal, logistical, and social morass to make it work. Aaron slept in a broken-down UPS van while he scouted the boundaries of the resort, and convinced the state to give him a business loan. He had to fight through a convoluted permitting process, and the town's controlling localism. Everything was more expensive and harder than they thought it would be, and it took years to get any locals on their side. In 2002 it became the first new ski area in Colorado in twenty years, and the first one built since ground was broken in 1982 at Beaver Creek, its polar opposite.

The Brills predicted the shift in skiing culture toward the backcountry. At the hill you have to carry avalanche gear, the terrain is unruly, and you work for your turns. Silverton, with its single lift and lack of infrastructure, is the antithesis of something like Vail's widespread conglomeration. It's refreshing in its simplicity and realness. But that's not to say it's without struggles. There are still locals who chafe at the Brills' desire to push boundaries. Plus, insurance is expensive, consensus is hard, and sometimes the Brills say that they'd sell because it's too stressful to run a ski area, but they're still here—a skier's mountain in a skier's town.

Aaron is notoriously taciturn, but Jen is chatty and filter free. At the table she shoots the conversation in a million different directions, and keeps getting up to talk with other tables. Shocklee, a long-term guide and lifelong dirtbag, is her opposite. He has a low-toned drawl, a quick smile, and a sneaky sense of sarcasm. He lives an ascetic dirtbag life, farther out on the edge than anyone else I've met who has truly made a life of it. To the point that he's become a kind of poster boy for ski bumming. There was a film made about how he's the last true skid, which features an MTV Cribs–style tour of his bare-bones 215-square-foot cabin in town. "Personal hygiene corner, don't need much room for that," he deadpans, as the camera pans over the stacked crates where he keeps his clothes and food. The film is called *A Fairy Tale*.

He was one of the Brills' first employees. He showed up to help dig lift towers the summer before they opened. He'd blown up from Durango on a whim, because it felt like town was getting too crowded. He heard that some yahoos were trying to start a ski area up Cement Creek, so he went to check it out. Jen and Aaron hired him on after he stuck around for a couple of days. He says it didn't hurt that his dog got along with their dogs.

He became a guide, and in the nearly two decades since he's spent winter shepherding skiers around the mountain, and summers in the Grand Canyon, guiding there, too, switching back and forth between seasons. He bought a river dory this year and he says that felt like a big step toward commitment. He's in his fifties, but he says he feels like he's in his twenties, and that in skiing he's found the fountain of youth. The quick grin behind his bushy, gray-tinged beard supports that, as do his snappy, fluid ski turns. But one of the reasons why Shocklee has been lauded as a folk hero is because he is rare. If his life is Jeff's stove metaphor, it's clear which burner he has on high.

When we're down to the last inch of beer in our glasses, I ask them the ski bum question: Do they think you can sustain it here in Silverton? They both say they've found a way to eke out a life here, but even Shocklee, the consummate forever skid, has some qualms about forever. The pipes in his cabin are currently frozen, and that battle against the elements feels exhausting sometimes.

We part ways with the promise to ski tomorrow, zip our hoods tight and head out in opposite directions down the wide, quiet streets. When I look back, Shocklee in his huge puffy is nearly invisible after a couple of steps in the silver swirl of the storm.

The winding gravel road up to Silverton Mountain follows Cement Creek, where the rocks are tea-colored from the dregs of old, abandoned mines. When you get to the point where you think the canyon can't get narrower, the lift appears out of nowhere on your right. It's an old slow double chair brought in after Mammoth Mountain in California retired it, nothing fancy or new, and it's emblematic of Silverton. You have to ski with a guide, except for a few days at the beginning and end of the season, and they only let about eighty people on the hill each day. But that exclusivity is also part of the appeal. The snow is generally untracked, because skiers are carefully spread out across the terrain. And even with a guide, a day of skiing there is less than the window ticket price of a day at Vail. Silverton's staff do significant avalanche control, to manage the touchy southern Colorado snowpack and stay a step ahead of skiers, but other than that they leave the mountain alone.

When I get there, guides are milling around the base, making groups and gearing up for the day. The base area consists of the lift, a ski rack, a couple of picnic tables, and the shuttle bus, which reads Silverton Mountain Correctional Facility on its side. I wave to Shocklee and stick my skis in the rack. Across the creek and up the hill, there's the wall tent that serves as a lodge, and an old broken-down school bus that houses the rental shop. Inside the tent there are disintegrating couches, a fire, a few T-shirts you can buy, and a mill of people signing paperwork and buckling their boots. Everything smells a little bit like wood smoke and worn layers.

I get sorted into a group with a father and daughter from Denver, a few middle-aged guys, a couple who work at Wolf Creek ski area, and one of the guide's girlfriends, a nurse in New Mexico who comes up as much as possible. We practice with our avalanche beacons, and then jump on the chair to the top. The goal at Silverton is to race through as many laps as possible. The runs are long and hard, and you're often hiking for turns, burning daylight and leg strength. The chair brings you about two-thirds of the way up the mountain, and from there you traverse or walk uphill. We immediately start boot packing up the ridge toward the billboard that marks the far end of the area's terrain. Courtney Walton, our guide, has to quickly gauge our skill and fitness to decide where he can take us, and hiking straight out of the gate gives him clues about how easily we move through the mountains. He watches to see if we stumble in the skittery loose black rocks, or get thrown off by the wind. Partway up the ridge he stops us, and tells us to carefully click into our gear, while he tests the snowpack. We drop into a chute called Rope Dee Four, which opens into a wide shoulder. Our first run is a glory lap of untouched snow, creamy and cold. The sky is a white swirl, and I get waves of vertigo in the featureless, open slope. I try to soften my knees, to trust the slope and to give in to gravity. Partway down I find a groove and let inertia carry me. When we regroup down at the base of the run everyone is giggling.

From there we go. No stopping for lunch or pee breaks. Scarfing down granola bars on the lift. One of the older guys peels off before the last run, but the rest of us are giddy. It's snowing harder every time we ride the chair. My body is made of noodles but I never want to stop.

"This is Neverland but the lost boys are a little more grown up," Krista, one of the new—and few female—guides, says about Silverton, when she comes to take a lap with us. And it does sort of feel like that. There is heavy responsibility here, and the consequences of these mountains are real. You need to be focused and educated to be a guide, but following someone like Shocklee also comes with a sense of immortality and escapism. If he's Peter Pan we are the lost boys falling in line, transfixed. It feels like I want skiing to feel, even though my lungs explode. Everything is glowy and full of delight. The whooping, sweeping downhills, sure, but also the tent, the beers afterward, the guys who helped me get my car unstuck. It feels like a place I could get lost in for a while.

Courtney's family has mined nearby for decades—his great-grandmother operated the Tomboy mine near Telluride. He loves being here, but he's also back in school to diversify his options, and some of the other guides mention alternate plans: nursing school, teaching, trying to have kids. Not everyone can be Shocklee. But maybe that's OK.

The night before, at the Avalanche, Jen told me that she had something to give me, and that I should stop by her office at the end of the day, so after I come down from the mountain I swing by, wobbly on my Jell-O legs. In a dusty room of the Greene Street Victorian that serves as their in-town office she digs up a cardboard box of skinny books and pulls one out for me. *Deep Powder Snow* by Dolores LaChapelle. I'd been trying to track it down for months. The 100-page booklets are $150 on Amazon and impossible to find anywhere else. I kept willfully hoping to uncover one in the back of a used bookstore, or at the library free box in some ski town, as a sign that I was searching out the right thing, but I'd been striking out. And now, apropos of almost nothing, Jen is pushing one into my hands, free, just because she feels like I need it.

Jen had mentioned Dolores in passing the night before, and once I have the book in my hands she tells me a little bit more about their relationship. Dolores was one of the pioneers of powder skiing, and the guru of a skiing-specific kind of philosophy, which she called deep ecology. It's the idea that skiing could connect us to the earth, and to a sense of purpose, and that moving downhill, in balance with gravity, was a way to understand our place in the ecosystem. Skiing, for Dolores, was a framework for how to live. She moved to the San Juans with her husband, Ed, an avalanche forecaster, in 1973. Previously they'd spent a long stint at Alta where they'd been among the first people to ski steep slopes in the canyon. Dolores was around when the Brills moved to town and she and Jen had become friends because they recognized a stubborn, independent, physical streak in one another.

When Dolores died in 2007, Jen bought up as many of the palm-sized paperbacks as she could find, to try to hold on to Dolores's vision of skiing as connection, and to make sure it got to the right people. I bumbled through a thank-you, touched, and I took the book back to Jeff's house, which was quiet. He was out working on the road, and the other guys were still in class. I stoked up the stove and carefully cracked the undersized book.

It's a full-on ski bum benediction. A rundown of the obsession, from her fascination with the big peaks of the Wasatch, to her early days ski instructing at Aspen (which she calls a pre-commune commune), to the way she finally settled into the San Juans, where she could practice what she called "tuning." It's the idea that when you're skiing, you're perfectly connected to nature. It's that feeling I got in the Silverton whiteout. When I relaxed and I listened, I could float.

Dolores, with her long braid, wild ways, and perfect turns, feels easy to retroactively deem an icon. Alta was new when she got there—she was the first person to ski Baldy Chute, and few skiers were in the San Juans when she and Ed, who helped invent the first avalanche beacon, moved down there. Their constant move toward untouched snow seems unattainable now. Dolores had the time and freedom to think about why it felt good, and she had a mission. Tuning was her purpose.

"Powder skiing cannot merely be considered a metaphor for living, but rather skiing powder shows us how to live. If we insist on proceeding arrogantly in the narrowly human centered world of modern culture we will continue to not only destroy the earth's species, but the very water and air on which we depend to live." She wrote that in 1993, and she died before the climate crisis was considered a catastrophe, but deep ecology is the opposite of the Anthropocene, and the human-centered, technology-driven life we live in. As her vision becomes less accessible, it also becomes more necessary. When she talks about the crucial flow of skiing she drills into something I still struggle to articulate, even when I'm deep in the downhill flow state. I'm not Dolores. I know I can't be Dolores, but there is something clear and beautiful about a focused life, and about trying to understand what skiing teaches you about yourself.

Sometimes I spend hours scouring Zillow, or winding down the wormhole of Instagram trying to find the perfect long-term life in the mountains: a community that's solid, a place I can afford, easy access to the mountains. I'm dealing with my own rush toward middle age and the idea that we might be the last generation to know the meaning of deep winter, and I want a way to feel all of that—to not let life slip by, and to feel tuned in. It doesn't have to be epic or even that exciting. I don't need the stomach clench of adrenaline every day, I just want the balance, a place and a passion. It's a more mature desire than what I dreamed up when I drove west the first time, but it still pushes against the limits of reality.

On the road, I live somewhere between elation and an ache, that bittersweet feeling of looking back, and trying to see the ways my life could have played out if I'd let it. Sometimes I still feel like I'm chasing a ghost of my former self. And even though I know where the myth breaks down, Silverton still kind of breaks my heart with longing.

The next fall I will get a message from Riley Lemm, who has moved out of his tent at Big Sky and is back in school. He is trying to figure out a way to work the system so he can ski all the time and still feel satisfied. "So do you think there are still ski bums?" he asks. "What did you find?"

What do I tell a twenty-two-year-old? No pressure, but you're it? Your guess is as good as mine? Take the good parts and run? We can't live up to the story we created, we're on the down swing, but it's still beautiful.

When I try to parse out the best parts of Silverton, I think back on Wendell Berry's idea about boomers and stickers, leavers and stayers, and why the connection matters. I've been a boomer so far, lit up by the search, untethered. Shocklee is a sticker with a booming heart. He dug in.

I'm still trying to come to terms with the fact that I didn't get everything I thought I would. I am trying to hold on to the places and things that I love. Trying not to wonder too hard what might have happened had I stuck. When I look at the people who are doing it: the Shocklees, and Guru Daves, and Groms who fit so perfectly into their environment, I know that they've committed—to the pain of loss, the ache of time, the joy.

My last morning in Silverton the sun comes out. The boys at Jeff's house are all busy, the road is still closed, and the mountain is booked out, so I'm stuck in town, alone and antsy. Avalanche danger is high, but to get some exercise, I decide to head up the very mellow slope above town, Shrine Hill, named after the Christ of the Mines shrine on its flank. I skin up alone as the day warms up, the coat of new snow glopping and sticking in the heat.

There are tracks ahead of me and I can see that one skier has ripped out a slide, right on the concave sunny side of the hill. There's inertia in the snowpack, and I know I could be an avalanche trigger if I'm not careful, so I'm cautious. At the top, when I re-layer, clip my helmet, and swig water, I waffle for a moment, second-guessing my plan, trying to decide if I've made the right choice. The town spreads out before me, and I can hear the bomb echoes in the distance, where Jeff is helping the DOT open the road.

I stick to my plan and ski the gradual, north-facing side of the hill, purposely avoiding the steeper, more dangerous slope, and the pull of the fall line. The snow is heavy and wet. It's changing underneath me, already not as good as it looked from the base of the hill, before I started, but a little bit of disappointment feels good in a messed-up way.

I am almost back to town when Jeff texts: We got the road open to the south, he says. You know you're welcome to stay, but this might be a good chance to get out.

I can't stay. I know that. By the time I get back to the house the next storm is already building behind the mountains. I load the car and swallow the bittersweetness of being on the road again. If you want it, there's always another mountain, another storm to chase, it's just a matter of how—and if—you choose to go on.

★ ★ ★ ★ ★

ACKNOWLEDGMENTS

This book exists because the ski world took me in and let me sleep on its couch. Literally and figuratively. In vaguely geographic order: Scotty and Forest, Molly and Drew. Alex and Todd. Travis and Marcus. Rachael Burks (I promise I wax my skis more now), Pat, JF, and everyone at the Chalet. TATS, whose conversations informed so much of this book. Mavis, Brad Kremske, Zarter, and the rest of the A-Basin crew. Adrienne Issacs, my eyes on the inside of the industry. Benny Wilson, Jeff Leger, Joe Paine. Bill Hoblitzell, for the dryer fact check. Riley Lemm, Cassie Able, Jesse and Sean (I know you were reading at night), MJ and Si guy. Miz Nicholson! Keely Kelleher, Sheldon Kerr, Mikey, Dave Powers, the Deep Powder House. Jeff Davis, John Shocklee. Jen Brill (I can't thank you enough for bringing me Dolores).

I relied on the wisdom of a range of academics, old timers, advocates, and sages. Thanks to Annie Coleman, whose book is invaluable. Liz Burakowski and the crew at POW for all they do. Auden Schendler, for his framework for making the world better. Timothy Tate, for soul. Ryan Burke, Sara Gibbs, Grom, the late John Fry, Greg DiTrinco, Michael Miracle, JK, CP, Kwame Harrison, Stan Evans, Art Findley.

I am lucky that my friends are also my heroes. Gratitude for the ASSFART brain trust, particularly Spencer, Sam Cox, Matt Hansen, John Howland, the Jaded, Sierra Shafer, Clare Menzel, and Britt Barnes (thanks for breaking into my car). Dave Reddick, for your vision in so many things.

Megan Michelson, Lyndsay Strange, Sally Francklyn, Katie Cruickshank (and Sunny!) for reading and giving me the guts to go on. Kate Schimel and Sarah Tory for digging through early drafts, and for being so good at both skiing and writing. Caitlin Littlefield for setting the timer to keep me accountable. Run club, for keeping me sane and semi-socialized while I finished a book in a pandemic.

Lady Shred, thank you for every trinity and every text, and for changing the ski world and the whole world. And so it shall be and even better.

Steph and KPL, see you in the hot tub. I told Springer to find us, I'm sure he will.

Stories and ski days don't mean much, book-wise, if you can't get them down, so I'm grateful to the good people at Aspen Words, who, among other things, put me up while I finished this manuscript. Space and time are the truest gift.

Zoe Sandler, who knew this story was mine before I did, and who has supported me in uncountable ways. Just let me know when you want to hit Plattekill.

The team at Hanover Square, particularly John Glynn, who seeded the idea and who expertly sliced through my fluffy language to make this book so much better. I am so lucky you got it, and get it.

And, most importantly, my people. My parents, who are unfailingly full of love and support, and who are the grown-ups I want to be when I grow up. And T, for reading it all, so many times, and for taking care of me as I work through it all: the words and the past and the future.

INDEX